MW00975681

Preston So

VOICE CONTENT AND USABILITY

MORE FROM A BOOK APART

Better Onboarding
Krystal Higginis

Sustainable Web Design
Tom Greenwood

Design for Cognitive Bias
David Dylan Thomas

Cross-Cultural Design
Senongo Akpem

Expressive Design Systems
Yesenia Perez-Cruz

Resilient Management
Lara Hogan

Everyday Information Architecture
Lisa Maria Marquis

Progressive Web Apps
Jason Grigsby

Flexible Typesetting
Tim Brown

Going Offline
Jeremy Keith

Visit abookapart.com for our full list of titles.

Publisher: Jeffrey Zeldman
Designer: Jason Santa Maria
Executive director: Katel LeDû
Managing editor: Lisa Maria Marquis
Editors: Sally Kerrigan, Caren Litherland
Technical editor: Mat Marquis
Book producer: Ron Bilodeau

ISBN: 978-1-952616-01-3

A Book Apart
New York, New York
http://abookapart.com

10 9 8 7 6 5 4 3 2 1

TABLE OF CONTENTS

For my father, with love.

FOREWORD

FOR EVERYTHING WE THINK comes naturally to us, we forget that someone, at some point, had to teach us how to do it. Acquiring natural language, for instance, requires years of frequent exposure, direction, correction, and interaction with other language-using humans. And any interface—whether just between humans, or between humans and machines—is only intuitive to the extent it resembles whatever an individual has learned how to use before.

Despite the hype, voice interactions with digital systems are not automatically easier to use than written websites. Only large investments of human effort come close to making it so. Creating good conversational content is wildly different from having a conversation. And as hard as it was for each of us to learn to speak, teaching a machine to do so is so much harder. You can't simply set a laptop down in front of Friends and hope it picks up the gist.

We need to deconstruct what we've forgotten someone taught us, reassemble sense from scratch (in all necessary permutations), and add human warmth and timbre without succumbing to our own comfortable biases about what sounding natural means.

No problem.

Fortunately, Preston So is here for you, drawing on his love of language as well as his direct experience with technology and content, to examine the emerging field of voice interaction and create a practical, principled guide to the task at hand. He provides clear steps to make voice content a possibility within any organization, because the process is going to be different for everyone. Soon, you'll be on your way to creating systems that are more accessible, inclusive, and intuitive for your audiences.

Teaching a computer to sound human is hard. Writing about it is harder. Preston has done both. Listen to him if you want to improve upon the silence.

—Erika Hall

INTRODUCTION

WHAT DO YOU PICTURE in your mind when you think of the word *content*? Or rather, what do you *hear*? Today, much of our content lives inside websites, splayed like wallpaper across browser viewports and smartphone screens.

But chances are that over the past year, you've interacted with at least a few, perhaps even dozens, of *voice interfaces*: experiences that serve users through aural and oral means rather than through written or printed media.

These days, voice interfaces are everywhere. We enlist our voices to schedule events on a calendar, plan a get-together, transfer funds between accounts, or order takeout for dinner. Voice interfaces add greater nuance and richness to our interactions with websites, phones, tablets, search engines, smart speakers, smart home systems, and Internet of Things (IoT) devices. Their use cases run the gamut, from manipulating switchboards for corporate phone hotlines to navigating websites with screen readers to teaching schoolchildren to read phonetically to staving off loneliness for elders and empty nesters.

Though they're rapidly becoming integral to our routines, today's voice interfaces are mostly limited to executing tasks and performing transactions on our behalf. But there are lots of other things we *could* do with voice interfaces for which we mostly still resort to the web. Demand is poised to intensify for designers, content strategists, and information architects to deliver *voice content*—richly structured information transmitted through the medium of voice—especially as the adoption of voice interfaces accelerates.

When it comes to delivering content—compelling copy, intriguing information, or just a dose of breaking news—voice interfaces still fall woefully behind. Outside of simple requests, users who want to use their voices instead of their screens to browse idly through a newspaper's articles, embark on a virtual audio tour of a museum's exhibits, or review a small business's product details are largely out of luck.

Voice content also remains mostly uncharted territory. As humans, interacting with one another by voice comes naturally to us, because speech is among our most primeval habits. By working with us on our own verbal terms, it's the machines that have to do what doesn't come naturally to them, along with parsing all the weirdness that characterizes human speech.

The growing interest in voice experiences and particularly in voice content puts in stark relief the other challenges surrounding the ongoing trend I call the *channel explosion*. Today, voice is just one specimen in a menagerie of new conduits for content (like augmented reality, digital signage, and IoT devices) that overturn our browser-friendly biases. For content practitioners and designers, our longstanding focus on the websites and applications we own—all largely visual and bound to devices with screens—will need to adapt to embrace other means of accessing content.

Working with voice content means handling content beyond the browser, a rapidly emerging reality for teams besieged by growing stakeholder demands. To be successful, we need to prepare our content for every conceivable channel under the sun while simultaneously investing the necessary time to make sure each individual experience—voice being just one of them—is the best it can possibly be.

Voice content is possibly the most removed from the content approaches we've long been accustomed to on the web. It means new workflows and new tools to navigate the journey from web content to voice content. Instead of long-form written content, we need succinct spoken content. Instead of visual design, we need verbal dialogues. And instead of visually rooted navbars, we need aurally rooted flows.

Voice content also scrambles all the neatly defined roles and responsibilities we used to treat as gospel. Because of its deeply interdisciplinary nature, everyone on the content or product team—be they designer, developer, copywriter, usability researcher, or accessibility specialist—needs to be involved in every step of a voice content implementation.

Despite its challenges, enabling voice content is an exciting emerging skill to add to any résumé. No two voice content projects look exactly the same, and there's no "right" sequence to follow since many stages overlap. In this book, we'll get solid footing for the work involved across the whole project lifecycle:

- In Chapter 1, we define what voice content is, what makes voice content voice content, why we should work on it in the first place, and how to get started.
- In Chapter 2, we repurpose existing web copy into voice-ready content by auditing it for its legibility in voice and acting on audit recommendations.
- In Chapter 3, we write our voice content into the elements of *dialogue*—prompts, intents, and responses—so it's understandable to voice interfaces *and* their users.
- In Chapter 4, we transform our dialogues into a *flow* by converting them into journeys in the form of call-flow diagrams, so voice content stays discoverable.
- In Chapter 5, we prepare our voice content for launch, conducting usability testing and prerelease testing, and completing other final steps before release.
- In Chapter 6, we cover the promising outlook ahead for voice content and discuss pressing issues of inclusion and representation.

Voice content is an outlier, but it's a thrilling one. Now is a great time to immerse yourself in it, because by witnessing your content buckling under the strain of so many disparate demands—with voice content requiring possibly the most exacting solutions of them all—you'll end up readier than ever for the future, with voice as just the first of many new and appealing ways to get to your content.

1 CONVERSATIONS WITH COMPUTERS

Conversation is not a new interface. It's the oldest interface.
—ERIKA HALL, *CONVERSATIONAL DESIGN*

WE'VE BEEN HAVING CONVERSATIONS for thousands of years. Whether to convey information, conduct transactions, or simply to check in on one another, people have yammered away, chattering and gesticulating, through spoken conversation for countless generations. Only in the last few millennia have we begun to commit our conversations to writing, and only in the last few decades have we begun to outsource them to the computer, a machine that shows much more affinity for written correspondence than for the slangy vagaries of spoken language.

Computers have trouble because between spoken and written language, speech is more primordial. To have successful conversations with us, machines must grapple with the messiness of human speech: the disfluencies and pauses, the gestures and body language, and the variations in word choice and spoken dialect that can stymie even the most carefully crafted human-computer interaction. In the human-to-human scenario, spoken language also has the privilege of face-to-face contact, where we can readily interpret nonverbal social cues.

In contrast, written language immediately concretizes as we commit it to record and retains usages long after they become obsolete in spoken communication (the salutation "To whom it may concern," for example), generating its own fossil record of outdated terms and phrases. Because it tends to be more consistent, polished, and formal, written text is fundamentally much easier for machines to parse and understand.

Spoken language has no such luxury. Besides the nonverbal cues that decorate conversations with emphasis and emotional context, there are also verbal cues and vocal behaviors that modulate conversation in nuanced ways: *how* something is said, not *what*. Whether rapid-fire, low-pitched, or high-decibel, whether sarcastic, stilted, or sighing, our spoken language conveys much more than the written word could ever muster. So when it comes to voice interfaces—the machines we conduct spoken conversations with—we face exciting challenges as designers and content strategists.

VOICE INTERACTIONS

We interact with voice interfaces for a variety of reasons, but according to Michael McTear, Zoraida Callejas, and David Griol in *The Conversational Interface*, those motivations by and large mirror the reasons we initiate conversations with other people, too (http://bkaprt.com/vcu36/01-01). Generally, we start up a conversation because:

- we need something done (such as a transaction),
- we want to know something (information of some sort), or
- we are social beings and want someone to talk to (conversation for conversation's sake).

These three categories—which I call *transactional, informational,* and *prosocial*—also characterize essentially every *voice interaction*: a single conversation from beginning to end that realizes some outcome for the user, starting with the voice interface's first greeting and ending with the user exiting the interface. Note here that a *conversation* in our human sense—a chat between people that leads to some result and lasts an arbitrary length of time—could encompass multiple transactional, informational, and prosocial voice interactions in succession. In other words, a voice interaction is a conversation, but a conversation is not necessarily a single voice interaction.

Purely *prosocial* conversations are more gimmicky than captivating in most voice interfaces, because machines don't yet have the capacity to *really* want to know how we're doing and to do the sort of glad-handing humans crave. There's also ongoing debate as to whether users actually prefer the sort of organic human conversation that begins with a prosocial voice interaction and shifts seamlessly into other types. In fact, in *Voice User Interface Design*, Michael Cohen, James Giangola, and Jennifer Balogh recommend sticking to users' expectations by mimicking how they interact with other voice interfaces rather than trying too hard to be human—potentially alienating them in the process (http://bkaprt.com/vcu36/01-01).

That leaves two genres of conversations we can have with one another that a voice interface can easily have with us, too: a *transactional* voice interaction realizing some outcome ("buy iced tea") and an *informational* voice interaction teaching us something new ("discuss a musical").

Transactional voice interactions

Unless you're tapping buttons on a food delivery app, you're generally having a conversation—and therefore a voice interaction—when you order a Hawaiian pizza with extra pineapple. Even when we walk up to the counter and place an order, the conversation quickly pivots from an initial smattering of neighborly small talk to the real mission at hand: ordering a pizza (generously topped with pineapple, as it should be).

Alison:	*Hey, how's it going?*
Burhan:	*Hi, welcome to Crust Deluxe! It's cold out there. How can I help you?*
Alison:	*Can I get a Hawaiian pizza with extra pineapple?*
Burhan:	*Sure, what size?*
Alison:	*Large.*

Burhan: *Anything else?*
Alison: *No thanks, that's it.*
Burhan: *Something to drink?*
Alison: *I'll have a bottle of Coke.*
Burhan: *You got it. That'll be $13.55 and about fifteen minutes.*

Each progressive disclosure in this *transactional* conversation reveals more and more of the desired outcome of the transaction: a service rendered or a product delivered. Transactional conversations have certain key traits: they're direct, to the point, and economical. They quickly dispense with pleasantries.

Informational voice interactions

Meanwhile, some conversations are primarily about obtaining information. Though Alison might visit Crust Deluxe with the sole purpose of placing an order, she might not actually want to walk out with a pizza at all. She might be just as interested in whether they serve halal or kosher dishes, gluten-free options, or something else. Here, though we again have a prosocial mini-conversation at the beginning to establish politeness, we're after much more.

Alison: *Hey, how's it going?*
Burhan: *Hi, welcome to Crust Deluxe! It's cold out there. How can I help you?*
Alison: *Can I ask a few questions?*
Burhan: *Of course! Go right ahead.*
Alison: *Do you have any halal options on the menu?*
Burhan: *Absolutely! We can make any pie halal by request. We also have lots of vegetarian, ovo-lacto, and vegan options. Are you thinking about any other dietary restrictions?*
Alison: *What about gluten-free pizzas?*
Burhan: *We can definitely do a gluten-free crust for you, no problem, for both our deep-dish and thin-crust pizzas. Anything else I can answer for you?*
Alison: *That's it for now. Good to know. Thanks!*
Burhan: *Anytime, come back soon!*

This is a very different dialogue. Here, the goal is to get a certain set of facts. *Informational* conversations are investigative quests for the truth—research expeditions to gather data, news, or facts. Voice interactions that are informational might be more long-winded than transactional conversations by necessity. Responses tend to be lengthier, more informative, and carefully communicated so the customer understands the key takeaways.

VOICE INTERFACES

At their core, *voice interfaces* employ speech to support users in reaching their goals. But simply because an interface has a voice component doesn't mean that every user interaction with it is mediated through voice. Because multimodal voice interfaces can lean on visual components like screens as crutches, we're most concerned in this book with *pure voice interfaces,* which depend entirely on spoken conversation, lack any visual component whatsoever, and are therefore much more nuanced and challenging to tackle.

Though voice interfaces have long been integral to the imagined future of humanity in science fiction, only recently have those lofty visions become fully realized in genuine voice interfaces.

Interactive voice response (IVR) systems

Though written conversational interfaces have been fixtures of computing for many decades, voice interfaces first emerged in the early 1990s with text-to-speech (TTS) dictation programs that recited written text aloud, as well as speech-enabled in-car systems that gave directions to a user-provided address. With the advent of *interactive voice response* (IVR) systems, intended as an alternative to overburdened customer service representatives, we became acquainted with the first true voice interfaces that engaged in authentic conversation.

IVR systems allowed organizations to reduce their reliance on call centers but soon became notorious for their clunkiness. Commonplace in the corporate world, these systems were primarily designed as metaphorical switchboards to guide customers to a real phone agent ("Say *Reservations* to book a flight or check an itinerary"); chances are you will enter a conversation with one when you call an airline or hotel conglomerate. Despite their functional issues and users' frustration with their inability to speak to an actual human right away, IVR systems proliferated in the early 1990s across a variety of industries (http://bkaprt.com/vcu36/01-02, PDF).

While IVR systems are great for highly repetitive, monotonous conversations that generally don't veer from a single format, they have a reputation for less scintillating conversation than we're used to in real life (or even in science fiction).

Screen readers

Parallel to the evolution of IVR systems was the invention of the *screen reader*, a tool that transcribes visual content into synthesized speech. For Blind or visually impaired website users, it's the predominant method of interacting with text, multimedia, or form elements. Screen readers represent perhaps the closest equivalent we have today to an out-of-the-box implementation of content delivered through voice.

Among the first screen readers known by that moniker was the Screen Reader for the BBC Micro and NEEC Portable developed by the Research Centre for the Education of the Visually Handicapped (RCEVH) at the University of Birmingham in 1986 (http://bkaprt.com/vcu36/01-03). That same year, Jim Thatcher created the first IBM Screen Reader for text-based computers, later recreated for computers with graphical user interfaces (GUIs) (http://bkaprt.com/vcu36/01-04).

With the rapid growth of the web in the 1990s, the demand for accessible tools for websites exploded. Thanks to the introduction of semantic HTML and especially ARIA roles beginning in 2008, screen readers started facilitating speedy interactions with web pages that ostensibly allow disabled users to traverse the page as an aural and temporal space rather than a visual and physical one. In other words, screen readers for the web "provide mechanisms that translate visual design constructs—proximity, proportion, etc.—into useful information," writes Aaron

Gustafson in *A List Apart*. "At least they do when documents are authored thoughtfully" (http://bkaprt.com/vcu36/01-05).

Though deeply instructive for voice interface designers, there's one significant problem with screen readers: they're difficult to use and unremittingly verbose. The visual structures of websites and web navigation don't translate well to screen readers, sometimes resulting in unwieldy pronouncements that name every manipulable HTML element and announce every formatting change. For many screen reader users, working with web-based interfaces exacts a cognitive toll.

In *Wired*, accessibility advocate and voice engineer Chris Maury considers why the screen reader experience is ill-suited to users relying on voice:

> *From the beginning, I hated the way that Screen Readers work. Why are they designed the way they are? It makes no sense to present information visually and then, and only then, translate that into audio. All of the time and energy that goes into creating the perfect user experience for an app is wasted, or even worse, adversely impacting the experience for blind users. (http://bkaprt.com/vcu36/01-06)*

In many cases, well-designed voice interfaces can speed users to their destination better than long-winded screen reader monologues. After all, visual interface users have the benefit of darting around the viewport freely to find information, ignoring areas irrelevant to them. Blind users, meanwhile, are obligated to listen to every utterance synthesized into speech and therefore prize brevity and efficiency. Disabled users who have long had no choice but to employ clunky screen readers may find that voice interfaces, particularly more modern voice assistants, offer a more streamlined experience.

Voice assistants

When we think of *voice assistants* (the subset of voice interfaces now commonplace in living rooms, smart homes, and offices), many of us immediately picture HAL from *2001: A Space Odyssey* or hear Majel Barrett's voice as the omniscient computer in *Star*

Trek. Voice assistants are akin to personal concierges that can answer questions, schedule appointments, conduct searches, and perform other common day-to-day tasks. And they're rapidly gaining more attention from accessibility advocates for their assistive potential.

Before the earliest IVR systems found success in the enterprise, Apple published a demonstration video in 1987 depicting the Knowledge Navigator, a voice assistant that could transcribe spoken words and recognize human speech to a great degree of accuracy. Then, in 2001, Tim Berners-Lee and others formulated their vision for a Semantic Web "agent" that would perform typical errands like "checking calendars, making appointments, and finding locations" (http://bkaprt.com/vcu36/01-07, behind paywall). It wasn't until 2011 that Apple's Siri finally entered the picture, making voice assistants a tangible reality for consumers.

Thanks to the plethora of voice assistants available today, there is considerable variation in how programmable and customizable certain voice assistants are over others (**FIG 1.1**). At one extreme, everything except vendor-provided features is locked down; for example, at the time of their release, the core functionality of Apple's Siri and Microsoft's Cortana couldn't be extended beyond their existing capabilities. Even today, it isn't possible to program Siri to perform arbitrary functions, because there's no means by which developers can interact with Siri at a low level, apart from predefined categories of tasks like sending messages, hailing rideshares, making restaurant reservations, and certain others.

At the opposite end of the spectrum, voice assistants like Amazon Alexa and Google Home offer a core foundation on which developers can build custom voice interfaces. For this reason, programmable voice assistants that lend themselves to customization and extensibility are becoming increasingly popular for developers who feel stifled by the limitations of Siri and Cortana. Amazon offers the Alexa Skills Kit, a developer framework for building custom voice interfaces for Amazon Alexa, while Google Home offers the ability to program arbitrary Google Assistant skills. Today, users can choose from among thousands of custom-built skills within both the Amazon Alexa and Google Assistant ecosystems.

Less programmable *More programmable*

Apple Siri

Amazon Alexa

Google Home

FIG 1.1: Voice assistants like Amazon Alexa and Google Home tend to be more programmable, and thus more flexible, than their counterpart Apple Siri.

As corporations like Amazon, Apple, Microsoft, and Google continue to stake their territory, they're also selling and open-sourcing an unprecedented array of tools and frameworks for designers and developers that aim to make building voice interfaces as easy as possible, even without code.

Often by necessity, voice assistants like Amazon Alexa tend to be *monochannel*—they're tightly coupled to a device and can't be accessed on a computer or smartphone instead. By contrast, many development platforms like Google's Dialogflow have introduced *omnichannel* capabilities so users can build a single conversational interface that then manifests as a voice interface, textual chatbot, and IVR system upon deployment. I don't prescribe any specific implementation approaches in this design-focused book, but in Chapter 4 we'll get into some of the implications these variables might have on the way you build out your design artifacts.

VOICE CONTENT

Simply put, *voice content* is content delivered through voice. To preserve what makes human conversation so compelling in the first place, voice content needs to be free-flowing and organic, contextless and concise—everything written content isn't.

Our world is replete with voice content in various forms: screen readers reciting website content, voice assistants rattling off a weather forecast, and automated phone hotline responses governed by IVR systems. In this book, we're most concerned with content delivered auditorily—not as an option, but as a necessity.

For many of us, our first foray into informational voice interfaces will be to deliver content to users. There's only one problem: any content we already have isn't in any way ready for this new habitat. So how do we make the content trapped on our websites more conversational? And how do we write new copy that lends itself to voice interactions?

Lately, we've begun slicing and dicing our content in unprecedented ways. Websites are, in many respects, colossal vaults of what I call *macrocontent*: lengthy prose that can extend for infinitely scrollable miles in a browser window, like microfilm viewers of newspaper archives. Back in 2002, well before the present-day ubiquity of voice assistants, technologist Anil Dash defined *microcontent* as permalinked pieces of content that stay legible regardless of environment, such as email or text messages:

> *A day's weather forcast [sic], the arrival and departure times for an airplane flight, an abstract from a long publication, or a single instant message can all be examples of microcontent. (http://bkaprt.com/vcu36/01-08)*

I'd update Dash's definition of microcontent to include all examples of bite-sized content that go well beyond written communiqués. After all, today we encounter microcontent in interfaces where a small snippet of copy is displayed alone, unmoored from the browser, like a textbot confirmation of a restaurant reservation. Microcontent offers the best opportunity to gauge how your content can be stretched to the very edges of its capabilities, informing delivery channels both established and novel.

As microcontent, voice content is unique because it's an example of how content is experienced in *time* rather than in *space*. We can glance at a digital sign underground for an instant and know when the next train is arriving, but voice interfaces hold our attention captive for periods of time that we can't easily escape or skip, something screen reader users are all too familiar with.

Because microcontent is fundamentally made up of isolated blobs with no relation to the channels where they'll eventually end up, we need to ensure that our microcontent truly performs well as voice content—and that means focusing on the two most important traits of robust voice content: *voice content legibility* and *voice content discoverability*. Fundamentally, the legibility and discoverability of our voice content both have to do with how voice content manifests in perceived time and space.

Voice content legibility

When we design how content will display in web environments, we often think about content *legibility* from the standpoint of text formatting and content flow. We calibrate our leading and kerning pixel-perfectly and wax poetic about typefaces and paragraphs. In turn, we debate whether a heading should use a particular color and whether an overwhelming wall of text should be split into more bite-sized paragraphs.

All of these considerations go out the window in the voice context. When it comes to voice content legibility, we're concerned about three things:

- **Context.** Voice content is necessarily contextless, because there's no way for a voice interface to know what preexisting knowledge the user has and, in most cases, what areas of the interface the user is familiar with based on previous interactions.

Web content
Highly contextual macrocontent

Voice content
Contextless microcontent

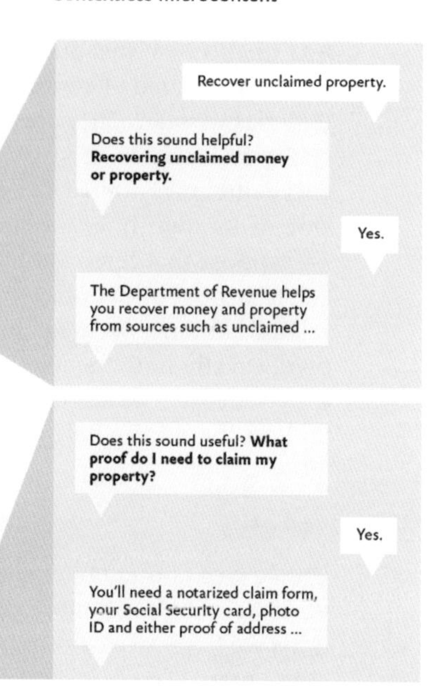

FIG 1.2: Web content benefits from the privilege of visual context, which voice interfaces lack. A frequently asked question (FAQ) could appear in the context of an FAQ page or on a credit card statement page. Voice content has no such luxury.

- **Verbosity.** Voice content is necessarily concise, because our attention spans and our levels of patience with aural interfaces are much more limited than with visual interfaces.
- **Spoken style.** Voice content is more colloquial and informal than the literary, formal web content we often find on websites.

Context

Voice experiences suffer from one key drawback: users have to assess content spoken by the computer in isolation from any context, a far cry from the highly context-laden content that browser experiences offer. In context-rich websites, for instance, we can immediately place ourselves in a topical context thanks to visual cues like headings and layout. But voice interfaces devoid of context need to work harder to help users understand where they are and what context they're in (**FIG 1.2**).

Every instance of voice content—every snippet of copy, every utterance—must therefore not rely on any context outside of its own internal context. Because we can no longer presuppose that certain content will coexist with other content (such as post teasers in a blog archive or articles in a list of related content), we need to read each piece of content on its own, exactly how the user in a real-world voice setting would experience it. The last thing we want our users to ask is: "How do I get there from here?"

Verbosity

Unlike web content liberally spread across browser viewports, voice content bears more resemblance to the quick conversations we have on a day-to-day basis at the checkout counter or momentary glances at digital signs. In other words, our content must reflect how we interact with our channels.

While the web emphasizes visual readability, voice, in contrast, forces us to treat legibility as aural listenability instead. Just as web users have little patience for walls of text, voice interface users bristle at wordy, overwrought prose. That means every instance of voice content must be efficient and to the point to mirror the brevity of typical utterances in normal human conversation. But voice legibility isn't just about patience; it's about retention, too.

Most American English speakers know what to expect after dialing a phone hotline. "Please listen carefully as our options have recently changed. Press one for reservations; press two for cancellations. *Para español, oprima número tres.*" The motivation for this monotone boilerplate has to do with how poorly we retain spoken language, despite our relative knack for recalling written language. Because utterances are about as permanent in our working memory as a ripple across a pond, voice interfaces need to do extra legwork to be concise so users can retain key information.

Verbosity tolerance in web and voice content—how long content can be before we start zoning out—is inversely proportional to how easily we can retain that copy. The tighter your content is, the easier it is to recollect later, and the friendlier it is for voice interfaces.

Depending on your audience's verbosity tolerance, you'll definitely want to optimize your content for brevity—use a stopwatch to time it if you have to!—as this has an outsize influence on how your users will judge your voice interface. Does your content remain just as "listenable" in a spoken setting as it does on the web in textual form? Are responses generated by the interface concise enough to ensure that users' minds don't wander?

Spoken style

Writing voice content has the unenviable distinction of relying on synthesized speech to convey meaning to the user. This places more pressure on interface designers to stay faithful to the cadence of normal conversation through a conversational tone and style. Authoring voice content is therefore more complex than other copywriting, as we react very differently to hearing something than we do to seeing it written out. Every instance of voice content must sound like a believable verbal contribution to a real conversation.

What makes a piece of voice content sound like authentic speech may be different for disparate audiences. Like a chat at the deli counter, informal speech in English tends to be slangy, with more contractions, y'know? Meanwhile, formal speech may take on an academic tone, like a Nobel lecture. Calibrating your voice content's internal style is a highly interdisciplinary task that requires careful attention to how your audience inflects their speech themselves. It's always best to cater voice content to the expectations of how your users prefer to hear it.

Just as weird approximations of the human body can lead to the uncanny-valley effect in robots, so too can the mechanical nature of synthesized speech make users feel awkward. Because users may already be distrustful, a voice interface displaying overfamiliarity can come across as creepy or threatening. As UX writer Joscelin Cooper has noted, even if interface text faithfully echoes the "lilt, flow, and syntax of human speech," there is always a risk of creating a "false intimacy that distances even as it attempts to foster familiarity" (http://bkaprt.com/vcu36/01-09).

Voice content also benefits from the added dimension of sound—if the technology supports it, information can be conveyed in a steady monotone, a breathy whisper, or a barely audible murmur. In voice content, this spectrum of sound mirrors the sort of style and formatting that HTML elements like `strong` and `em` articulate for the web. The best voice content makes optimal use of the human traits and sonic qualities natural speech has to offer, encoding additional context and subtext rather than reciting everything in a monotonous drone.

But good voice content isn't just about being legible for users interacting through voice. Because voice content lacks any tether whatsoever to other contexts, good voice content also has to be reachable and discoverable, not orphaned or siloed. After all, what use is making a piece of content more concise and spoken in style if we can't even get to it in the first place?

Voice content discoverability

No matter how legible you've made your voice content, it's only going to be useful to users if they're able to find it when they need it. That's where *discoverability* comes into play: how *reachable* our voice content is once we enter the interface.

A typical website might guide users to an exhaustive sitemap, or simply bring a compelling related piece of content to the current page and position it somewhere prominent. But in voice interfaces, every piece of voice content must be aurally revealed; there's no way to build efficient sitemaps. Delays can quickly compound as a result, as users struggle not only to understand the interface's overall structure but also the pathways to get to what they need. This makes discoverability an especially pressing concern for voice content.

When it comes to the discoverability of our voice content, we're concerned about three things:

- **Depth.** Every additional step deeper into the interface saps just a little more of the user's attention. Thus, every important item of voice content must be available as readily in the interface as possible for quick discovery.
- **Duration.** Discoverability is also about the length of time it takes to arrive at a desired piece of content. Therefore, every important item of content must be accessible as quickly as possible regardless of its structural depth.
- **Reachability.** If an item of voice content is part of your interface, it should be reachable by some mechanism even if it isn't the most important piece of content in the mix.

Depth

"A few clicks away" on a website might translate to "a few steps away" on a voice interface, but the two could not be more different from each other. Whereas clicking on a sitemap link requires no more than a few milliseconds, following steps to a piece of voice content might require answering multiple questions in succession from the voice interface.

Voice interfaces, because they deploy events in time rather than documents in space, hold users captive whenever they speak. Each time a voice interface has to deliver information or confirm something a user said, the user's destination dangles ever farther away. Limiting the number of back-and-forths to get to needed information therefore reduces user frustration.

To evaluate voice content discoverability for its depth, we can measure the number of steps required to arrive at each individual content item (or, at minimum, the most important ones). For simpler experiences, a suitable benchmark might be discovery of a content item within two to five steps. For more complex interfaces, a depth of eight to ten steps might be more appropriate.

Duration

Depth and duration are two distinct metrics, since the amount of time it takes to get to a destination might bear no relation to the number of steps it takes to get there. Even if you can limit the number of back-and-forth exchanges with the voice interface to get to desired content, the journeys might take far too long, whether due to the verbosity of each utterance by the interface or because the user is having trouble responding to a confusing question.

Measuring the duration of a journey from an initial entry point to a desired piece of content is a matter of timing the number of seconds it takes to get there. For simpler interfaces, a quick trip in and out within thirty seconds might suffice. But for interfaces with a bit more sprawl or long-windedness, durations of one or two minutes might be a more appropriate benchmark to aim for.

Just as with voice content legibility, the ideal depths and durations for your voice content depend on the value your audience attaches to certain content and the use cases you're serving. A voice interface providing information about area restaurants will likely have a lower threshold for both characteristics than a rather more complex voice-driven encyclopedia.

Reachability

Another critical concern is ensuring that no individual piece of voice content is orphaned or unreachable. It can be easy to let less essential items of voice content fall by the wayside, beyond reach. If your voice interface will use search, results should be as unambiguously named as possible so they're not accidentally demoted into oblivion. And if you're juggling a large quantity of content, reachability can be a particularly pressing concern.

Evaluating a piece of content's discoverability in the context of reachability is a straightforward yes-or-no question: Can I get to every piece of voice content from the starting point of the interface, somehow, some way?

Our voice interfaces should ensure that users can discover each and every piece of voice content without having to understand the full hierarchical structure of the interface from the get-go. That means limiting depth and duration and maximizing reachability.

WHAT MAKES GOOD VOICE CONTENT?

The social conversations we have on a daily basis on the street corner or at the watercooler are fundamentally different from the informational content we access through voice interfaces. But certain characteristics straddle both our spoken conversations and what we call voice content.

Untethered to any visual or physical context, voice content is necessarily contextless—short and sweet—and conversational, just like the idle chat we might have with our favorite person at the local deli about what's new on the menu.

Voice content also has to be discoverable. Users get more frustrated the longer or deeper a journey takes them, and the ultimate frustration is an unreachable or orphaned piece of content.

Now that we grasp the underpinnings of what makes good voice content, we can turn our attention to a much harder problem: how to extricate the content cloistered in our websites and transform it into voice-ready content.

2 GETTING CONTENT READY FOR VOICE

IN MID-2016, at DrupalCon New Orleans, I sat down to catch up over lunch with my dear friend Nikhil Deshpande, chief digital officer of the state of Georgia. Over sandwiches, we discussed Deshpande's ambitious vision for a better state resident experience, powered by voice interactions, that would weave its way through the structured content on the Georgia.gov website.

Most voice interfaces in the mid-2010s were primarily transactional rather than informational, operating in realms like pizza delivery and credit cards. Soon after the launch of the Alexa Skills Kit—a tool for developers to create custom-built *skills*, or apps that add functionality beyond Alexa's core capabilities—we decided to design and build an experimental voice interface capable of answering any Georgian's question about state government. The goal of the now-decommissioned Ask GeorgiaGov Alexa skill—the predecessor to the current text-driven Ask GeorgiaGov chatbot—was to find a way to reduce the load on agency phone hotlines and limit unnecessary in-person visits by serving Georgians in the comfort of their own homes (http://bkaprt.com/vcu36/02-01).

Not only would it be among the first Alexa skills to deliver content; it would also be the first-ever Alexa skill built for residents of Georgia. With my Acquia Labs colleague Chris Hamper, we kicked off our quest to trawl for convenient content on Georgia.gov that we could repurpose into a voice content corpus.

As we started looking, however, we realized we had only scratched the surface of our moonshot mission's complexity. The groundbreaking content strategy the Digital Services Georgia team had executed with such success for their website wasn't going to translate so easily to managing voice content alongside web content. Keeping one version of content would certainly be more maintainable, but it would serve voice interfaces imperfectly at best. Alternatively, developing a voice-specific version in parallel with web copy would mean more overhead for the Georgia team, despite the benefits of optimizing separate versions for web and voice.

In a lengthy heart-to-heart with the Digital Services Georgia team, we discussed this conundrum. Given that the team lacked the resources to create fresh verbal content from the get-go, we realized we'd instead need to build Ask GeorgiaGov with existing copy.

This meant repurposing the crisp, serifed text on Georgia.gov as synthesized speech, a format profoundly alien to its original context and worlds away from the comforting confines of a visual browser. We needed to excise ourselves from the web-only context and consider the ramifications of recasting Georgia.gov's tightly curated copy as voice content.

WHEN TO WRITE, WHEN TO REUSE

Chances are good that you already have a slew of content that can help you develop a voice content experience. Even if the content you're creating is distinct from your existing brand content, it's important to take stock of what you already have. You may need to have an honest chat with stakeholders eager to author new content, because introducing another version of channel-specific content, this time for voice, can lead to content siloes and maintenance nightmares further down the road.

There are pros to writing new content versus reusing existing content. Creating new content means you can have:

- **More control over how "conversational" and "spoken" your voice content sounds.** Existing content, because it's used in other channels, needs to stay largely the same (unless a full revamp of all content is up for debate), meaning it will never quite fit voice interfaces hand in glove.
- **Flexibility to structure and plan content however you want for long-term maintenance.** Existing content is generally imprisoned in a content management system or some other rigid schema that doesn't allow for large-scale changes after new channels (like voice) are introduced.
- **The ability to distinguish voice content through voice-specific advantages.** Writing net-new voice content allows designers to take advantage of all the benefits of the voice medium. For example, sound gestures like klaxons that indicate an error are more appropriate for aural interfaces.

That said, reusing existing content for voice in lieu of writing net-new content also comes with significant advantages from a content strategy perspective. Reusing existing content opens the door to the following benefits:

- **Content reuse means only worrying about the glue between pieces of content.** Net-new voice content might demonstrate the sort of conversational or spoken style that voice content should, but the surrounding glue—how users are brought to their voice content—is just as important in voice interfaces. If you reuse existing content, the only new writing you'll need to do is craft the interface text that bridges gaps between individual content items. That's half the work of writing all new content, where you'd have to write copy and also the glue between it.
- **Content reuse leverages existing editorial processes and structures.** Most organizations already have clear editorial processes and workflows in place, as well as procedures for how content goes live. Reusing existing content means you won't have to integrate voice-specific content into those

processes or conceive new ones to handle voice-only content—saving steps in either case.

- **The ability to manage a single cross-channel rendition of content.** Reusing existing content and maintaining a single version of your content means you'll always have parity between your web content and voice content. This means you'll be able to maintain accuracy and brand consistency by managing just one version of content across all channels rather than across multiple versions at once, which are always at risk of drifting out of sync.

Though reusing web content is generally the better course of action, that doesn't mean you should treat your web copy as gospel. In *Voice Interaction Design*, voice designer Randy Harris argues that websites currently representing the primary entry point for information seekers should be considered merely another "interface" rather than the "primary" way of accessing content (http://bkaprt.com/vcu36/02-02). "Voice enabling the Web [*sic*] for the sake of voice enabling the Web is pointless," Harris writes, "but there are web sites [*sic*] galore with promise for useful speech interaction." In other words, consider how your content shows a web-first bias and how to work toward a more channel-agnostic approach that doesn't privilege one channel over another.

The voice of voice content

What's the "voice" you intend users to hear when they interact with your content? Is it that of a hotel concierge, or an airport information desk worker, or an investor-relations manager? Or is it more like the voice of a close confidant or front-of-house staff at a beloved restaurant?

When we design interfaces for voice content, we need to consider not only the spoken language inherent to voice content but also the natural inclination of humans to apply identities and personalities to their voice interfaces.

The vast majority of visual and physical interfaces around us—computers and televisions, kiosks and devices—seldom impose a human identity or brand voice on a user experience.

But synthesized speech is different, because humans use it to infer ideas about the sort of person they're speaking with, whether those inferences are by accident or by design.

Brand consistency is a big issue for organizations, because new content for voice could clash with other channels if it displays contradictory traits, like a goofy sense of humor when the home page copy is serious and measured. Though it's important to inject personality into new voice content to showcase your brand voice, it's always possible to go too far. Be careful with humor and other potential sources of creepiness that will drive users away instead of attracting them.

Because they approximate human identities more than any other class of interface, voice interfaces need to consider how new voice content affects the existing brand voice and brand consistency across all channels. Voice content can be more flexible than written text, but it still needs to bear some relation to already existing copy.

Whether you're adding to your content or reusing existing content, remember that voice content is characterized by its *legibility* (contextless, concise, and in a spoken style) and its *discoverability* (depth, duration, and reachability). Here are some questions you can ask yourself as you evaluate both new and recycled content:

- How will this content match up with other copy representing the brand experience, like that found on the website or other channels? Does it drown it out or send mixed messages?
- How will this content borrow from current content to preserve brand consistency?
- How will you plan and structure this content so it makes sense in a larger context for copywriters and editorial teams? And how will it stay current and in sync with updates applied to other channel-specific content, if any?
- How will approval processes and editorial workflows evolve in an environment where content destined for multiple channels needs to be previewed and approved?
- How will issues like regulatory compliance and legal requirements change your existing processes and workflows now that there are multiple channels of content?

- As you can see, regardless of whether you've opted to write new voice content or repurpose existing content, your content strategy and processes still require ample discussion with your stakeholders and a review of your existing content.

The challenges of content creation

It's already difficult to manage content for a single website, let alone an array of channels that couldn't differ more from one another. Add competing editorial prerogatives to the mix, and things can get even more volatile. After all, many organizations have only recently transitioned to web-driven copy as part of their migrations from analog to digital. And many of them also have cash-strapped editorial and engineering teams that can only focus on what's most important.

If you have the luxury of kicking off some new content writing, you're one of the lucky ones. Voice content requires a very different sort of writing from the more literary, formal registers common on the web. But deciding whether to write new copy or repurpose existing copy also poses urgent questions for content teams, because there isn't a one-size-fits-all answer for all organizations.

For this reason, explore reusing the existing brand content you have before considering an entirely new version of content just for voice, especially when it comes to sensitive content that has to stay up to date. Reworking existing content means you get to take advantage of your content in multiple ways—it pops up on your website, your voice interface, and anywhere else you can think of. That sort of content reuse is where we turn our attention next.

VOICE-FRIENDLY CONTENT

If you've opted to repurpose your existing content so it does double duty as voice content, your next step is to gather together content that is already somewhat conversational. We call this content *voice-friendly content*, because though it might be stylistically conversational in nature, it still needs some rejiggering to be ready for prime time in a voice interface.

Much web content remains littered with browser-dependent turns of phrase (like "in the footer" or "on the previous page") or calls to action (like "read more" or "click here") that are nonsensical to a voice user. After all, one of the issues with reusing websites as content sources for voice interfaces is the fact that, especially for teams with small budgets, preexisting web content is by and large still optimized for web scenarios rather than contextless experiences.

And though many CMSs are excellent for maintaining *structured content*—content planned and developed in a modular way so it can be displayed in almost any channel—the vast majority still retain a bias for web content over other manifestations of content by overemphasizing how content appears as HTML. In other words, even if you're using a CMS that purports to manage content for presentations beyond the browser, it's likely you're still writing copy contextually suited for the web, not for voice.

Identifying voice-friendly content

Because some of our web content is often already written with a conversational cadence, we can often find at least some preexisting content that lends itself well to voice interfaces. To find voice-friendly content, take a look at the current state of your content with an eye toward copy that's already structured like a conversation:

- **Frequently asked questions (FAQs).** FAQs are experiencing something of a renaissance. Content strategist Caroline Roberts calls them "satisfying and efficient" in how they can easily morph into a question-and-answer machine (http://bkaprt.com/vcu36/02-03). Because they're written in a more conversational style to begin with, FAQs are excellent examples of voice-friendly content. Voice interfaces serving FAQ content generally provide a search tool as an entry point for users to find a desired topic or question.

- **Instructional content.** Content formatted as recipes or step-by-step tutorials or procedures also lends itself well to voice content, because each step is generally finite and needs action on the user's part to proceed. These short outlays of instructions can be forklifted into a voice interface with little overhead. Voice content consisting of instructions tends to display *progressive disclosure,* a model in which incremental tidbits of information are given out in chunks, rather than in one fell swoop as a tidal wave of content, so as not to overwhelm the user.
- **Multiple-step forms.** Just like content realized in a step-by-step cadence, forms that require multiple steps to serve relevant content, such as customer support, feedback, or registration forms, are also excellent candidates for voice content due to the need to collect and return small amounts of information at crucial moments. Wizard-like forms that help users get to specialized information as quickly as possible are also great candidates, such as a voice-driven COVID-19 self-survey that delineates what steps to take after feeling symptoms.

To find voice-friendly content on Georgia.gov, we pored over available copy and ended up focusing on the most conversational section of the website: the FAQ pages. Compared to other areas of the Georgia.gov site, which were much more expository in nature, the FAQs were already concise, free of a lot of extra context, and written in an informal style.

The Georgia.gov website's "Popular Topics" section contains pages with key information and relevant FAQs for each subject (**FIG. 2.1**). But we had one big problem: because Georgia's editorial team understandably wanted only one managed version of content (not one for web and another for voice), we realized that many of the FAQs, now extricated from their web-only trappings, would perform poorly if recited by a voice assistant in isolation.

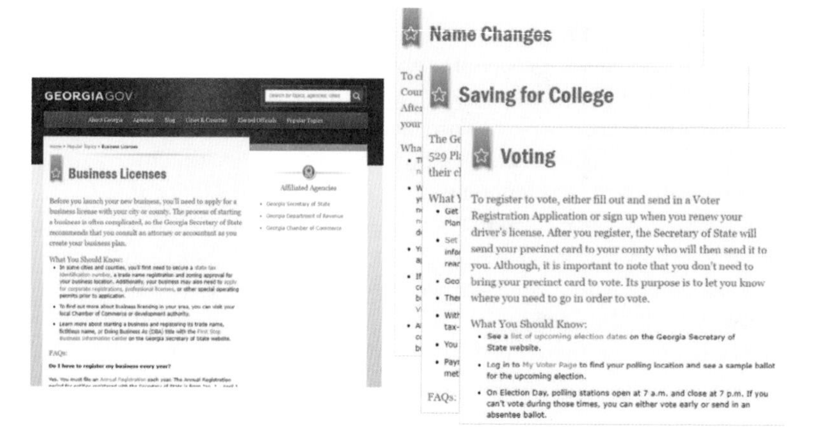

FIG 2.1: This series of screenshots from Georgia.gov's "Popular Topics" section (before its 2020 redesign) shows how Georgians could access information about business licenses, name changes, saving for college, and voting on distinct pages that contained an introduction, a "What You Should Know" section, and a series of FAQs.

Deciding what problems to solve

Here's a head-scratcher: You can't really pinpoint the right mix of user problems you can reasonably solve with voice content if you haven't already trawled through your web content to find the voice-friendly material—but then, how will you know what you're looking for if you don't go into that quest knowing what problems you're trying to solve? Generally speaking, these steps need to happen in parallel due to how much they inform each other.

We knew from our initial discussions with the Digital Services Georgia team that we wanted to answer Georgians' questions, but *how* exactly that would occur remained in flux until we had our full slate of voice-friendly content. To inform our requirements, we looked closely at how users came across information on the Georgia.gov site through search queries, and especially at how the voice-friendly content in the "Popular Topics" section was structured.

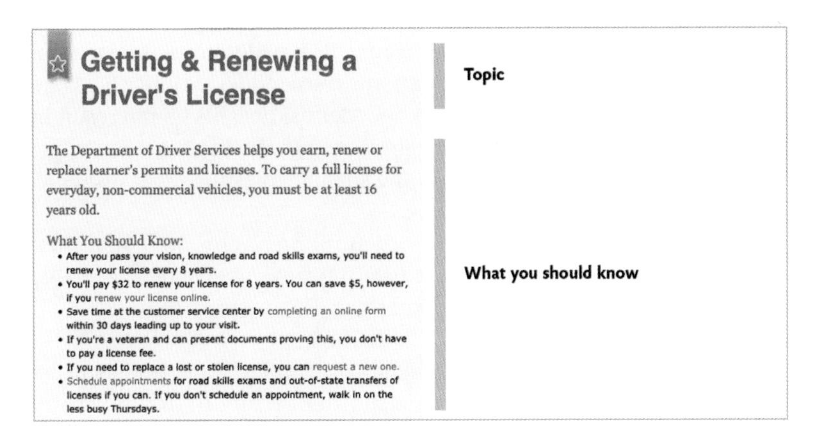

FIG 2.2: In this example, the "Driver's Licenses" topic page consists of the topic header, an introductory paragraph, and a "What You Should Know" section, which together form an overarching summary of the topic.

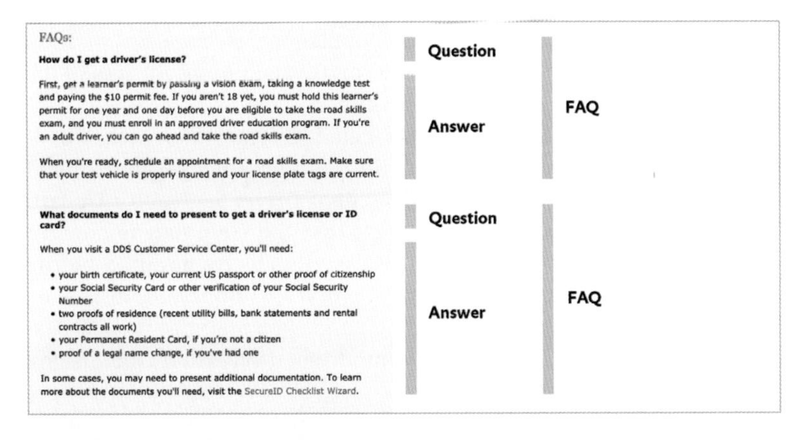

FIG 2.3: A typical question-and-answer couplet making up a single frequently asked question (FAQ).

We noted that there was a clear and obvious content hierarchy on Georgia.gov. Each "Popular Topics" page was made up of two major sections: a "What You Should Know" section with general information about the topic, and an "FAQs" section that dug deeper into that topic (**FIG 2.2–3**).

Just as a web user would consult the high-level information at the top of the page first, we resolved to forklift the same prioritization scheme into our Alexa skill. This would result in progressive disclosure, beginning with the most general information and proceeding to the specific questions that residents might have about the topic.

By understanding the structural hierarchy, we could then begin developing the core user journeys we wanted to enable for our Ask GeorgiaGov interface. We began with two:

- ask a question about a state government topic (and let users choose the right search result) and receive initial "What You Should Know" information
- select a desired FAQ from a list of available questions and receive a readout of the FAQ answer

These user journeys clearly follow the hierarchical structure of the content itself on each "Popular Topics" page. But we soon realized our initial scope had omitted an important area of the website's content we had missed.

Before 2020, Georgia.gov included a sidebar on each "Popular Topics" page containing links to relevant state government agencies such as the Department of Driver Services or Department of Revenue. This meant that the Georgia.gov content within our scope wasn't necessarily always going to be the end of the journey—in fact, a completely different state agency website outside the remit of Georgia.gov might be the real destination a Georgian has in mind when accessing Georgia.gov content through a browser.

To maintain functional parity between the website and the voice interface, we enabled users to benefit from those linkages to state agencies having their own websites (and phone hotlines) by introducing a third journey to allow users to acquire an agency phone number:

- request an agency phone number that a user can call for unanswered questions

This provided a clear analogue to the agency link on the website sidebar.

As this story illustrates, it's important to work through user requirements and content development in parallel. Until we understood how users were making use of the existing web copy, we didn't have a full enough picture to solidify all user requirements, much less the overarching mission.

Getting a sense of the problem scope

Once you have a relatively sure-footed idea of the problems you want to solve for your users and the voice-friendly content you want to deliver to them to honor those objectives, it's time to get your hands dirty.

Equipped with our user journeys and voice-friendly content from Georgia.gov, we examined how each topic page would translate into the voice user experience. We wanted to get some sense of how close our voice-friendly content was to being ready for an interface with no visual component.

Imagining the voice-friendly content we identified in the context of a real-world voice interface immediately revealed risks. Asking a question about a topic would trigger a search across all of the names of the "Popular Topics" pages, meaning that each topic title needed to be unambiguous. Once the topic was selected, Ask GeorgiaGov would introduce the subject by reciting the page's introductory paragraph and the "What You Should Know" section, with the option for users to dig deeper with FAQs farther down the content item.

For this step, we had several meetings with the Digital Services Georgia team to discuss some of the ramifications of the risks we found. In the process, we sat down with the editorial, user experience, and engineering teams to share our conclusion that some reworking of content would be required. User experience designer Rachel Hart, who graciously offered us deep insight as a stakeholder into how Georgians discover and consume their content, detailed why disambiguation between similar-sounding topics and FAQs was so essential:

If you say "No" to an FAQ question, Alexa skips to the next FAQ, and the next, until you say something sounds helpful or Alexa runs out of questions. When the user chooses what they want to hear, they need to know exactly what they're committing to. We need to make sure that our labelling—for both titles and FAQs—is clear. (http://bkaprt.com/vcu36/02-03)

Fortunately, each question-and-answer couplet was structured in a conversational format already. But what about the language within, prose that was to be spoken by Alexa but was never originally meant to be recited aloud except in screen readers?

Just because a piece of content is *voice-friendly* doesn't mean it's *voice-ready*. To ensure the voice-friendly content we gathered for our identified problems was voice-ready, we needed to calibrate our existing content so it could perform optimally in voice interfaces.

AUDIT FOR VOICE-READINESS

To make voice-friendly content more voice-ready, we can turn to a *content audit*, which might not have been the first thing to jump to mind in this context. Content audits aren't just for web content strategists and compliance officers; they're also useful for weighing how your ostensibly channel-agnostic copy performs across voice and other conduits like augmented reality, digital signage, and smart televisions.

Content audits examining voice content delivery need to pinpoint foibles that don't pose problems on a typical website but will trip up your voice user. Auditing content requires us to untether ourselves from long-trusted editorial elements like links, calls to action, and contextual references.

In short, a voice content audit focuses on how content *sounds* in isolation and offers insight into how to optimize it for delivery through voice interfaces.

Prepare your questionnaire

Once you have a rough sense of what would make your voice-friendly content voice-ready, it's time to prepare an audit questionnaire that you can repeat for each content item. This means sifting through and inspecting every content item under a microscope to gauge its suitability for a voice interface. For Ask GeorgiaGov, we wanted to identify those tricky puzzles for users caused by links, calls to action, and other browser-based elements. In the process of reading every topic page and FAQ aloud, we asked ourselves three questions:

- Where does each link in a content item (in this case, an FAQ response) lead, and does the link make sense in isolation within the content item? (For example, if there is a link that says "Contact us" within the FAQ response, does that link make sense to a voice user or is it superfluous?)
- Are all content items (in this case, FAQ questions) unambiguous such that they don't require any additional context? (For example, is every question phrased in such a way that it makes sense in isolation, on its own?)
- Are there any issues with references to places outside the immediate content item? (For example, is there a reference to a different section on the page that would be less readily discoverable in voice?)

If your voice-friendly content is replete with web elements, your questionnaire might include additional questions:

- Does every instance of formatting or style make sense outside of the web environment? (Most voice interfaces lack the ability to interpret boldfaced, italicized, underlined, or otherwise formatted text.)
- Does every media asset (images, videos, audio files) have a voice-ready transcript or substitute such as alt text? (With the possible exception of audio files or audio-only video playback, voice interfaces cannot handle media.)

Title	No.	Recommendation	Proposed Accepted Y/N	Changes Made on Live Site Y/N
Getting Copies of Criminal Records (Georgia Crime Information Center)	2	fee structure sheet: **Proposed:** delete sentence, remove link (included later on this page already): "A fee, payable by certified check or money order to the Georgia Crime Information Center, is required." **Current:** "Reference the fee structure sheet for applicable fees."	Y	Y
		Federal Bureau of Investigation		
—How do I run a background check on an applicant for employment?	1	Read more about this: Delete sentence; use previous sentence. **Proposed:** "If, however, you'd like a fingerprint-based record check, you'll need to fill out a Georgia Crime information Center Service Agreement and complete an enrollment form for the Georgia Applicant Processing Service (GAPS). **Current:** "If, however, you'd like a fingerprint-based record check, you'll need to fill out a Georgia Crime Information Center Service Agreement and complete an enrollment form for the Georgia Applicant Processing Service (GAPS). Read more about this."	Y	Y
—How much does a background check cost?	1	fee structure sheet: **Proposed:** "Fees vary from $10 to $48.25 based on the type of services requested. A fee structure sheet *(PDF, 192 KB)* is available on the Georgia Bureau of Investigation website." **Current:** "Reference this fee structure sheet."	Y	Partially - added second sentence , waiting for approval from agency before adding first
Getting Your Tax Refund (Department of Revenue)	0	No action needed		
—How do I check the status of my refund?		**Question:** "How do I check the status of my tax refund?"	Y	Y
—What are a tax refund offset and an offset notice?	1	View a list of these agencies: Delete sentence; use previous. **Proposed:** "If you want more information about	Y	Y

FIG 2.4: A page from our content audit document cataloguing topic pages, FAQ questions, FAQ answers, and all links, along with whether the recommendation has been accepted and whether it has been published.

Legend

	Keep intact
	Reword surrounding sentence per suggestion (action-based link; no context)
	Rewrite phrasing of question; also needs further discussion
	Needs further discussion
	Other
	No action needed

FIG 2.5: The legend for our content audit document, with different colors indicating no change needed, modification recommended, and outliers needing discussion or refinement.

Don't be alarmed if your audit questionnaire needs to shift midway through the content audit. It's seldom the case that teams are able to home in on the precise mix of questions on the very first try. If you do make changes midway through, be sure to repeat the audit for those content items you've already evaluated so you don't miss a spot.

Read, record, recommend

Once your audit questionnaire is ready to go, it's time to begin jotting down each problem area and making a recommendation to resolve it. For Ask GeorgiaGov, we prepared a color-coded audit table that accounted for every topic page, FAQ question, and FAQ answer, along with every link therein (**FIG 2.4**). In the table, we classified recommendations according to a legend and tracked whether proposals were accepted and whether they were already implemented live (**FIG 2.5**).

WHAT TO AUDIT FOR

Even a brisk audit of your corpus of content can reveal ways in which your copy inadvertently stops voice audiences in their tracks. And even if you've found voice-friendly content like

FAQs that lend themselves to reuse in a voice interface, you need to grapple with some tough questions before allowing its reuse to proceed willy-nilly.

Auditing links and calls to action

The web is littered with characteristic emblems that take full advantage of the medium: links, buttons, calls to action, and media assets are all integral to how we interact with websites. We traverse links to find a new rabbit hole, follow calls to action to buy a product, and click play buttons to stream videos. In other visual environments like written chatbots, these tropes of the web can usually function normally.

In voice and other contextless experiences, however, these commonplace artifacts may in fact block the user's ability to interact with content in meaningful ways. Such vestigial features of the web—the link to nowhere, the unrealizable call to action, the unviewable image—make it harder to ensure that our previously web-biased content, teeming with links and calls to action, is amenable to voice users.

Links are potential recipes for disaster in voice content. After all, how would a voice interface click and follow a hyperlink to display a web destination? As Aaron Gustafson wrote at *A List Apart*:

> *Links that read generically like "click here" and "read more" are not terribly useful, especially when the text of every link is being read out to you—which is a key way headless UI users skim web pages. Make it clear where you are linking. Restructure sentences if you need to in order to provide good link text. (http://bkaprt.com/vcu36/01-05)*

Hyperlinks are problematic not solely because they refer to inaccessible islands; they also implicitly outline context for the content in question ("for more information") and moonlight as calls to action ("learn more"). In web content, links also depend heavily on their affordance, whereas in voice, there isn't a way to color text blue and underline it. Such artificial indicators disappear in synthesized speech, flattening links into their surrounding text so they melt away undifferentiated.

Calls to action in voice content also present challenges. A link to nowhere pales in comparison to a failure to follow an instruction expressly signaled by a call to action (such as "visit this website" or "read more below"). When there's no ability to click a button or access a website, as users would typically do in a browser environment, we can't prescribe impossible actions for them to perform in our voice content.

For Ask GeorgiaGov's FAQ responses, we worked hard to ensure that links and calls to action remained legible in a voice context without their typical telltale traits in a web context. We also recontextualized certain links and calls to action into surrounding prose so they would simultaneously allow voice users to proceed unimpeded and avoid obscuring the available action for web users:

BEFORE	AFTER
Medicaid has several programs that can help you and your family. Learn more about these programs.	Medicaid has **several programs** that can help you and your family.
For both the elderly and those with disabilities, waiver programs provide medical assistance within the comfort of a home- or community-based environment. Health professionals come to you to help you through recovery. Learn more about these programs.	For both the elderly and those with disabilities, **waiver programs** provide medical assistance within the comfort of a home- or community-based environment.
You can update the name on your passport by securely mailing your application and fee to the U.S. Department of State. Find out what you need to do before sending the application.	You can **update the name on your passport** by securely mailing your application and fee to the U.S. Department of State.

Auditing for voice legibility

Beyond links and calls to action, we also want to limit the need for *context* and *verbosity* when it comes to voice content. These factors, unique to contextless settings like voice, cannot be manipulated normally by physical and visual means.

Consider the fact that many web pages undertake a cumulative approach to information disclosure; the flow and hierarchical structure of a page both mirror how a user consumes it. At the top lies a bit of background information, while the deeper detail that follows presumes a user has already read what came beforehand. This helps the user add to their understanding with each incremental uptake of information.

In voice interfaces, however, these luxuries go out the window. If an FAQ answer farther down a page relies on prior contextual knowledge settled by a previous FAQ, or if an answer points to an "aforementioned" question or the "above" answer, the lack of context will confuse the user. Our web content is replete with these sorts of *phantom references*, rhetorically or visually interdependent signals that guide web users but are meaningless to the voice user audience.

In voice, we can never presume that users have "read" any content outside where they are. Phrases like "earlier on this page" and "previously" are out (and to become more channel-agnostic, we should avoid these sorts of contextual and locational references in web content from the get-go, too).

Since Georgia handles multiple kinds of advertisements, benefits, and loan applications, we modified the text of questions having to do with each of these, anticipating that users might run into problems disambiguating between questions handling unemployment benefits as opposed to disability benefits. Unambiguous topic names and questions helped make associated content more discoverable via search.

BEFORE	AFTER
How do I submit an advertisement? (on "Subscribing to the Farmers & Consumers Market Bulletin" page) ("*Which* advertisement?")	How do I submit an advertisement to the Farmers and Consumers Market Bulletin?
How long can I receive benefits? (on "Applying for Unemployment Benefits" page) ("*Which* benefits?")	How long can I receive unemployment benefits?
How do I start my loan application? (on "Applying for Environmental Loans and Tax Credits" page) ("For *which* loan?")	How do I start my environmental loan application?

Just as voice lacks the contextual cues of websites, voice content also shows a clear preference for concision. Voice interfaces that overdeliver on information will drive users away immediately.

Some modifications proposed by a content audit can do double duty when it comes to addressing verbosity. In this Ask GeorgiaGov example, rewriting nonsensical links and calls to action also allowed us to abridge the content so it would read more gracefully for voice users:

BEFORE	AFTER
The Georgia Lottery Corporation only raises money for the Georgia Pre-K and HOPE Scholarship programs; it does not administer those programs. Read more about the Pre-K Program and the HOPE Scholarship.	The Georgia Lottery Corporation only raises money for the **Georgia Pre-K and HOPE Scholarship** programs; it does not administer those programs.
Although the state provides a standard homestead exemption of $2,000 for your primary residence, some counties offer taxpayers even greater exemptions. Learn about these exemptions for your county.	Although the state provides a standard **homestead exemption** of $2,000 for your primary residence, **some counties offer taxpayers even greater exemptions**.

Auditing for cross-channel interactions

Though channel-specific content audits can do wonders to identify weak spots in our content, limiting your perspective to a single channel doesn't match reality. Users seldom stick to one device during their quest for information and may swap at will between a smartwatch, smartphone, tablet, and desktop computer during their search. A single *cross-channel interaction* may consist of a voice interaction and interactions with multiple other devices, like Alexa to iPhone to MacBook. The interactions might even be happening at the same time!

Although content audits should focus on the channel under investigation (in this case voice), considering at least some other possible channels can uncover scenarios where a different interface may be better for the purpose at hand. For instance, a user may feel stumped by a reference to an external website in voice content, but if they have a laptop open next to them, we should accommodate their desire to switch from Alexa to their laptop.

Be mindful that your attempts to keep website links and phantom references out of the picture don't limit a user more than expected. For example, if a user sees a link in website copy that isn't reproduced in the voice interface in some fashion—such as through a brief aside for listeners who happen to be next to a phone or computer—the overall value of the voice content might decrease. Setting a somewhat level playing field for channels or privileging one over the others is a question of trade-offs and how your users interact with your other channels.

During our audit of content for Ask GeorgiaGov, we opted to keep certain references to external government websites intact (instead of removing them, as we would have done at first due to their being beyond our initial scope) to oblige users who have a browser waiting across the room on an iPad or PC. Moreover, we included content sourced directly from the linked websites, which is managed by other editorial teams, to furnish even more context relevant to the information at hand—and to maximize the voice content's value even for those with other devices at the ready.

BEFORE	AFTER
See the differences between an LLC, Sole Proprietorship, and Corporation.	Not sure which business type meets your needs? Learn the differences between an **LLC, Sole Proprietorship, and Corporation** on the Georgia.gov blog.
Read about licensed practical nursing requirements and transferring a license from another state.	You'll follow a different application process to get your nursing license depending on if and where you've been licensed before. The Georgia Secretary of State website provides **licensed practical nursing requirements** and **licensed registered nursing requirements**.

In other instances, we overtly mentioned external contexts like the Georgia Lottery and Georgia Bureau of Investigation websites. This prevented the external resources from being shrouded in a nonfunctional call to action, since a listener would only hear the text.

After all, how would a voice user know that "Benefitting Georgia" content sits on the Georgia Lottery website? In cases like this, where it was clear that the external link was adding value to the information, we inserted needed context from the linked agency site so users could make an educated decision about whether to run over to a browser or sit tight with Alexa.

BEFORE	AFTER
Visit Benefitting Georgia to read about how much money winners claim, students receive in aid, and retailers earn for their businesses.	Visit **Benefitting Georgia** on the Georgia Lottery website to read about how much money winners claim, students receive in aid, and retailers earn for their business.
Reference this fee structure sheet.	Fees vary from $10 to $48.25 based on the type of services requested. A **fee structure sheet** is available on the Georgia Bureau of Investigation website.

That users often do in fact want to consult multiple devices illuminates one of the key critiques that content strategists face in today's growing milieu of digital experiences. Far too often, we pigeonhole users into standalone channels such as websites without acknowledging the multimodal manner by which digital natives move seamlessly, from device to device, to quench their thirst for relevant content.

Therefore, content audits should treat content as the free-form fluid that it truly is. *Content is like water*, to borrow an aphorism from responsive web design (http://bkaprt.com/vcu36/02-04). As it morphs into various shapes and sizes, content should mirror the multifaceted journeys users routinely take as they jump with nary a concern between wearables, smartphones, tablets, laptops, and, of course, voice interfaces.

MANAGING AUDIT RECOMMENDATIONS

Once your content audit is complete, you'll have an impartial account of the problem areas and recommendations to address them. Your team will need to evaluate each recommendation individually to determine whether it should be applied, modified, or ignored. For Ask GeorgiaGov, we reviewed each proposed revision one by one with the editorial team, honestly weighing the risks in the process and making sacrifices where necessary.

Many of these modifications are easier said than done. Any change to content invites debates with stakeholders. And no revision can take place without due consideration for both the original web context and other channels, since a change could hamstring web users or upend the multidevice journeys users embark on. The trade-offs can be severe, and the debate can be pitched, but it's a critical stepping-stone to a more agnostic content strategy.

It's always possible to go too far. Content strategists must ensure that the revisions don't reduce any value already infused into web content. In some cases, you may need to prioritize compliance with government regulations or adherence to a product requirement over keeping voice content short and to the point. And in other situations, subtle contextual references may still be necessary to ensure that web users still benefit from the advantages of browser mainstays like links.

As such, it's important to gauge each recommendation emerging from an audit with a clear-eyed perspective. Like revising the first draft of a book, it helps to spend substantial time away from your content audit before revisiting it so it's easier to stay impartial when discussing audit recommendations.

Every time you come across an identified problem area in your list of recommendations, such as a wayward link or an impossible call to action, evaluate the advice from the standpoint of one overarching question: "Are there any unintended consequences if we make this change?" This will help you identify not only the obvious potential ramifications but also those that were missed during the audit.

All of the questions you should ask yourself and the team with each recommendation relate to these unintended consequences:

- Is there a reduction in voice content legibility or discoverability? (Does the content become less readable or less reachable after this change?)
- Will it make cross-channel interactions more difficult or less efficient?
- Will it lower the accuracy, quality, or value of the content itself?
- Will it lower the accuracy, quality, or value of the content on the web? on mobile? on tablet? on a virtual reality headset? (If you only have one version of content across all channels, evaluate that content in the context of all other channels your content travels through.)
- Are there any other unintended consequences we didn't consider?

One of the most difficult tasks of evaluating audit recommendations is prioritization. Editorial stakeholders have responsibilities as varied as regulatory compliance, legal issues, and corporate branding. Because stakeholders from different backgrounds can offer input from varied perspectives, it's best to gather everyone in a single room and discuss each recommendation frankly and openly. Generally speaking, if a piece of content is problematic along multiple fronts or in more disruptive ways, that item should be addressed before others that might only show one symptom of voice-unreadiness.

Auditing content for voice-readiness can be a complex process, especially when it comes to reviewing recommendations with stakeholders. But it's rewarding work, because it's a great excuse to take a closer look at your content—something that can be all too rare in organizations where content strategy sometimes takes a back seat to other concerns. And content audits can support your forays into other channels as well, inching your content ever closer to a future beyond the web.

VOICE IS JUST ONE CHANNEL

Content audits assessing how consumable content is in novel settings aren't just limited to voice in terms of their value for content strategists. Digital signage and information kiosks, which prize readability at a distance, or smartwatches and smart televisions, which truncate lengthy copy, are excellent candidates for a content audit with an eye on context and verbosity.

How will the content you've now extricated from your website look and behave, or read and sound, in synthesized speech on a Google Home, in an 8-point font on a smartwatch app, in an 864-point font on a digital sign in a train station, or at a 15-degree angle projected through an augmented reality overlay?

Many of the same lessons we learned from our content audit for Ask GeorgiaGov carry over to content audits for other never-before-seen destinations for our copy. But it's important not to privilege one channel at the expense of others. A content audit that doesn't focus on just one possible user experience can spotlight areas where rich cross-channel possibilities go undetected, like stepping across the room to consult a laptop while still within earshot of Google Home.

Our content auditing experience with Ask GeorgiaGov proves that we must also audit content in a channel-agnostic way with due consideration for how users may rely on an assortment of interfaces in their search for information. As such, at least a somewhat cross-channel view is essential to ensure the viability of your content across discrete experiences.

Of course, though a content audit can reveal how users consume content on one or multiple devices, content strategists and editors are still responsible for the hard work of striking the correct balance when it comes to content beyond the web. As we watch our content continue to detach from the traditional website, it has never been more urgent to ready ourselves for a new paradigm of content with no assumptions about what is in a user's hands or in their range of view.

Now that our content is ready for voice, we can begin to design—or rather, write—its nesting grounds, its load-bearing trusses: the dialogue.

3 CRAFTING DIALOGUES

AT ITS CORE, voice content is made up of two key elements. *Dialogues,* which we'll cover in this chapter, are the templates into which we pour our voice content. Every conversation we have with someone else, human or machine, can be converted into writing. Dialogues represent the aural equivalent of the "body" of a web page, as well as the searches, clicks, and scrolls we use to navigate to the content we want on a website.

Meanwhile, *call flows* are the structures that guide interaction. We'll take a closer look at these in Chapter 4, but for now, all we need to know is that in voice interface design, the language of dialogue and the structure of call flows are highly interdependent.

To create a cohesive voice interface, designers, information architects, content strategists, and developers must work closely in concert. "Crafting the dialogue and the call flow are parallel in the sense that they occur simultaneously; neither comes first," writes Harris in *Voice Interaction Design* (http://bkaprt.com/vcu36/02-02).

In other words, neither can be designed in isolation from the other. Dialogues move users forward and take them to their desired content through verbal exchanges, while call flows govern how they get there structurally. Let's walk through dialogue to begin.

THE ELEMENTS OF DIALOGUE

Each word, sound, and silence is a design element in voice content. Instead of a visual canvas, we have an empty soundscape to fill with audio. Instead of an eye for design, we need an ear and a tongue for well-spoken, understandable, and, most importantly, actionable language. For this, we need to deploy the *dialogue elements* that inflect the interplay between human and voice interface.

For transactional voice interactions, this tends to mean a call-and-response paradigm that is typical of a back-and-forth conversation, because utterances by the voice interface tend to be shorter. For informational voice interactions, however, the arrangement of dialogue elements becomes much more complex and nuanced, because the length of utterances can be more unpredictable.

As humans, we have an innate understanding of how dialogue works. Machines, on the other hand, need structures that mirror the internal workings of how they process and serve information. Every voice interface, and thus every dialogue, deals with four primary dialogue elements:

- onboarding
- prompts
- intents
- responses

Onboarding: Orienting the user

Every conversation with a voice interface begins with making a good first impression. The *onboarding* step is critical, because it settles the voice interface's identity, encourages trust, conveys value, and suggests a clear next step—all within a few short moments. Onboarding involves both introducing the voice interface to the user and orienting the user around an initial next step.

Because voice interfaces tend to have a user's undivided attention during the first few seconds of a conversation, this is the optimal time to communicate purpose and first steps. For this reason, conversational designer Amir Shevat argues in *Designing Bots* that the onboarding stage is essential in order to "generate a trigger that will build a usage habit" and justify its value proposition (http://bkaprt.com/vcu36/03-01).

The more you can stay to the point with a brief greeting and welcome, the quicker the user can begin to unlock voice content. So how can we kick off a voice interaction successfully? Let's start with the first impression given by Ask GeorgiaGov upon introducing itself:

Ask GeorgiaGov: ***Welcome to GeorgiaGov.***

Ask GeorgiaGov establishes its identity quickly, its purpose implicit in its short and sweet self-introduction. Formal but not aloof, Ask GeorgiaGov presents itself as a friendly and helpful government official. The onboarding is as brisk as possible, at the slight risk of proceeding almost too quickly.

Your onboarding should aim to be as clear as possible about the motivations for using the voice interface, especially if users are thrust into a voice interaction without any context. We were able to keep the onboarding short for Ask GeorgiaGov in part because we knew that its initial users would be coming from the Alexa marketplace, which displays a comprehensive description of the Ask GeorgiaGov skill ahead of its installation. Given the chance to repeat the project, we'd consider providing some additional background for those who may not have installed the skill on their Alexa device themselves:

Ask GeorgiaGov: *Welcome to GeorgiaGov,* ***your source for state government information.***

Remember, however, that the additional descriptiveness of the self-introduction might not be desirable. There's always a tipping point past which more onboarding becomes less valuable and more verbose to the user's ears, delaying the fulfillment of their goals. Imagine, for example, if the bot kept going:

> *Ask GeorgiaGov:* *Welcome to GeorgiaGov,* ***your source for state government information. I can answer any question you have about state government in Georgia.***

For designers, it's important to consider the trade-offs of front-loading more information for a more prescriptive voice interaction—a form of hand-holding—versus allowing the user to proceed unencumbered and choose their own adventure.

Prompts: Eliciting user responses

At the end of the onboarding, voice interfaces should offer an immediate next step. This initial suggestion signals what sort of response the user should give, and also begins to shape the user's mental picture of the interface's internal structure.

Prompts are dialogue elements whose purpose is to elicit a response from the user. Much like a call to action on a website, prompts are the primary way voice interfaces can glean how a user wants to proceed.

For Ask GeorgiaGov, after onboarding the user, the voice interface prompts the user to take an action.

> *Ask GeorgiaGov:* *Welcome to GeorgiaGov.* ***Feel free to ask a question.***

Here, the user is asked for the first time to respond to the interface. Until now, it has primarily been the interface's show, demonstrating what purpose it serves and why it can be trusted. Now it's the user's turn to share what goal they're trying to achieve.

Prompt design has a significant influence on a voice user's experience, not solely because it's the first thing users respond to, but also because it's usually the last thing users will hear each time the interface speaks. How prompts are worded, and how often they are cemented in the user's mind, will dictate much of the subjective experience users have with your interface.

Prompts can be *non-directive*, as we see with "Feel free to ask a question": an open-ended prompt that invites a response entirely improvised by the user. But non-directive prompts can often be so unrestricted that they offer the user very little actionable information about how to meaningfully kick off an interaction.

A better approach would have been a *directive* prompt, explicitly giving the user a template to adhere to as they respond:

Ask GeorgiaGov: *Welcome to GeorgiaGov. Feel free to ask a question,* ***like "Who is the governor of Georgia?" or "How do I register to vote?"***

Indeed, in The Conversational Interface, the authors note that directive prompts tend to be far more successful than non-directive prompts (http://bkaprt.com/vcu36/01-01).

If you use directive prompts, minimize the cognitive load on users by making the specific words needed by the interface the very last thing users hear before they respond. This sort of recency helps users keep their options top of mind, according to Cohen, Giangola, and Balogh in *Voice User Interface Design* (http://bkaprt.com/vcu36/01-02). It's particularly important in voice interfaces because we humans tend not to retain speech well.

Prompting with menus in voice interfaces

If users aren't presented with a directive prompt containing binary options or seeking a wildcard response, chances are they need some form of menu to account for multiple choices. Unlike web menus and navbars, voice menus are problematic; not only do they need to be brief, but the hierarchies they reflect can't be nearly as complex as drop-down menus on large websites.

Spoken menus are ideal for users who want to move rapidly to the content they're looking for, but overused and oversized menus can overwhelm users. They're "easy to browse visually" but "painful to review" in a voice interface, writes Cathy Pearl in *Designing Voice User Interfaces* (http://bkaprt.com/vcu36/03-02). As McTear, Callejas, and Griol state in *The Conversational Interface,* voice interface designers must make a choice: they can either present longer menus at the risk of losing the user's attention, or present menus more often at the risk of annoying the user (http://bkaprt.com/vcu36/01-01).

To ensure users can find their way to their desired content, limit menu size to three or four items, and pass additional options to a "second page" to keep machine utterances succinct and to parcel out less frequently selected choices. "Unfolding in time means holding people captive," writes Harris in *Voice Interaction Design,* and "there are no progress bars in voice interfaces" (http://bkaprt.com/vcu36/02-02). He suggests notifying users when a lengthy list is in the offing and either indicating how many items are in the list or asking the user to confirm they're ready before proceeding.

Intent: Understanding what the user wants

In order to respond to the interface's opening prompt, the user also needs to formulate an *intent*—the motivation to do something, whether it's to say yes to a question or to find out how to renew a driver's license.

If a prompt elicits voice input from users, then intent represents the underlying rationale bringing people to the interface in the first place. Though intent is a common term in user experience circles, particularly when it comes to mapping user intent through an interface, it's imbued with much more meaning in voice, because a user's expression of their intent in various forms is the only way for devices to serve responses catering to users' needs.

When approaching Ask GeorgiaGov, the user may have an intent of learning how to renew their driver's license and can express that in response to the interface's prompt:

Ask GeorgiaGov:	*Welcome to GeorgiaGov. Feel free to ask a question.*
User:	***How do I renew my driver's license?***

Intent identification, the process by which voice interfaces tease out a machine-understandable input from a human response, is hard. It requires voice interfaces to comprehend the user's expressed intent, which may not be clear or in scope. People seldom say things in immediately machine-understandable ways.

Once Ask GeorgiaGov hears the user pose a kickoff question—their first intent—in response to the initial prompt, it works to match it with the most likely result from a search triggered across Georgia.gov's content:

Ask GeorgiaGov:	*Welcome to GeorgiaGov. Feel free to ask a question.*
User:	*How do I renew my driver's license?*
Ask GeorgiaGov:	***Does this sound helpful? Driver's licenses.***
User:	*Yes.*

Here, Ask GeorgiaGov is stating that it interprets the user's question "How do I renew my driver's license?" as a wish to access the topic "driver's licenses." Note how the interface issues a new prompt ("Does this sound helpful?") in response to the user's initial question in order to confirm that it correctly identified the intent.

Since the prompt is a yes-or-no question, once Ask GeorgiaGov hears "yes" in response to its intent identification, it considers the identified intent correct and proceeds to deliver the voice content associated with the intent. If the user answers "no," Ask GeorgiaGov knows that the intent remains unidentified and continues down the list of search results until the user says "yes."

Ask GeorgiaGov:	*Welcome to GeorgiaGov. Feel free to ask a question.*
User:	*How do I renew my license?*

Ask GeorgiaGov: ***Does this sound helpful? Getting a boating, fishing, or hunting license.***

User: *No.*

Ask GeorgiaGov: ***Does this sound helpful? Driver's licenses.***

User: *Yes.*

Because humans are fickle creatures, prone to shroud their truest intentions, voice interfaces often need to deduce what a user's intent is based on limited information. Though we can delegate many of these difficult inferences to the voice assistants or development platforms we depend on, some voice interfaces could use a little extra help.

Intent identification with slots and tokens

For many prompts, *intent* is simply the user's response to a prompt in its entirety, like in the Ask GeorgiaGov example where the user's expressed intent is matched with the term "driver's licenses" through a direct search of the query "How do I renew my driver's license?" in the content database. In this case, the user's response is forklifted wholesale into the search engine as the user's intent, which Ask GeorgiaGov then confirms it understood correctly by providing search results and asking, "Does this sound helpful?"

Sometimes, however, voice interfaces need additional insight, particularly for open-ended prompts (like the classic wildcard "How can I help you?") that accept a range of possible responses in multiple formats. In interpreting these arbitrary user intents, they may need to slice and dice a user's response to tease out words or phrases that are important parameters for intent identification and jettison those that don't matter as much. That's where *slots* and *tokens* come into play.

Where an explicit mapping of arbitrary user input to clear machine input needs to take place, voice interface designers can explicitly indicate *slots* that represent where a particular parameter understandable to machines, a *token*, is expected.

Tokens can be as extensive as words and phrases—like the city of Alpharetta or Fulton County in Georgia—or as tiny as numbers or an interrogative tone. Knowing whether a user has asked a question or is merely making a statement is often an essential part of intent identification as well.

To illustrate this, let's briefly consider a hypothetical scenario in which Ask GeorgiaGov asks a user's location to better serve them with county-specific information:

Ask GeorgiaGov:	*Welcome to GeorgiaGov.* ***Where do you live in Georgia?***
User:	*I'm currently in Alpharetta.*
Ask GeorgiaGov:	***Is this correct? Fulton County.***
User:	*Yes.*
Ask GeorgiaGov:	*Thank you. Feel free to ask a question.*
User:	*How do I renew my driver's license?*

Let's unpack what happened here. If the user answered with the city of Alpharetta in the Atlanta suburbs, how did Ask GeorgiaGov infer Fulton County, the county in which Alpharetta is located? The answer has to do with how intent identification is handled differently between the prompt asking for a location ("Where do you live in Georgia?") and the prompt asking for a search query ("Feel free to ask a question.").

"How do I renew my driver's license?" doesn't need any further intent identification, because it makes up the entire search query. "I'm currently in Alpharetta," however, requires the voice interface to compare chunks of what the user says to a database of cities and counties and then offer a confirmation prompt ("Is this correct? Fulton County.") in order to later filter out information that's irrelevant to Fulton County residents. If Ask GeorgiaGov accepted "I'm currently in Alpharetta" at face value without identifying a matching county, this intent-identification process would result in an error.

Writing dialogues and reading them aloud can help your team identify where more focused intent identification is necessary to match discrete pieces of user input to a single cohesive machine input. This form of intent identification—sifting through the user's response and picking out what's important—is often the most custom-built portion of an interface when it comes to development.

Let's take a look at another hypothetical example that follows the same progression and illustrates how interfaces can either outline *explicit slots,* which follow a strict, unchanging pattern, or *implicit slots,* which are inferred from a user's open-ended response.

Ask GeorgiaGov:	*Welcome to GeorgiaGov.* ***Where do you live in Georgia?***
User:	*I live in Macon, but I work in Warner Robins.*
Ask GeorgiaGov:	***Is this correct? Bibb County.***
User:	*Yes.*
Ask GeorgiaGov:	*Thank you. Feel free to ask a question.*
User:	*How do I renew my driver's license?*

Macon is located in Bibb County, whereas Warner Robins is located primarily in Houston County. The user has offered two separate locations that map to different counties in Ask GeorgiaGov's data set, even though Ask GeorgiaGov only prompted the user for their residence, not their workplace. So how would Ask GeorgiaGov know whether to use Macon or Warner Robins as the user's expressed intent?

In this case, the user has given Ask GeorgiaGov a couple of clues by saying "I live in Macon" *and* "I work in Warner Robins," but the interface needs to do a bit of extra work to understand which is more relevant to the user's intent.

Many voice interfaces obligate the user to respond according to a set of predefined templates, like "I live in **[city or county]**,"

"My home is in **[city or county]**," simply "**[city or county]**," or, as we saw earlier, "I'm currently in **[city or county]**." These boilerplates outline explicit slots where the user inserts a city like "Macon" or a county like "Bibb County" as the token. Because "I work in **[city or county]**" isn't one of the templates provided by the voice interface for explicit slots, the voice interface ignores this portion of the utterance.

But using explicit slots can prove to be overly brittle, since there's always a chance a user will respond outside the scope of available templates. Saying "Macon's home" or "I hail from Macon"—when "**[city or county]**'s home" and "I hail from **[city or county]**" aren't formats the voice interface is equipped to handle—leads to an out-of-domain error. This is why teams using explicit slots need to do much more due diligence in thinking through how users might respond.

Voice interfaces that have stronger capabilities for interpreting spoken language will instead parse the user's utterance and make intelligent guesses about the most important information contained inside. Implicit slots interpreting the utterance "I live in Macon, but I work in Warner Robins" will choose one of the two tokens ("Macon" or "Warner Robins") to prioritize for intent identification. Because implicit slots require deeper understanding by the voice interface, they're often more difficult to implement.

To us as designers, the distinction between explicit and implicit slots may seem like an immaterial implementation detail, because, ultimately, both approaches ought to yield the same result. Whether defined explicitly or implicitly as a slot, the interface correctly pinpoints the user's domicile, not workplace, as the right token to use when identifying the surrounding county as the machine input.

Nevertheless, we need to account for these possible journeys together with other stakeholders, especially our architects and developers, because this sort of intent identification also increases complexity. For simplicity's sake, Ask GeorgiaGov avoids parsing slots within user utterances, preferring to interpret the full utterance—the search query it passes to Georgia.gov—as the intent instead.

Voice interfaces are still fundamentally machines; they're closed systems that require understandable input to generate output. For informational voice interfaces and voice content, slots might accept tokens such as search terms, keywords, taxonomy terms, or complete questions. But accepting a wider range of responses in open-ended prompts means more potential points of failure. Unless you're confident you can pull it off technically, sometimes it's best to keep things simple by merely treating the full utterance as gospel—to take users at their words.

Responses: Delivering information

Responses are how voice interfaces deliver critical feedback, customer support, and, of course, voice content. Whether it's a confirmation or an error, the answer to a question, or the result of a search, responses are essential for voice interfaces, where feedback delivery is limited to aural and verbal cues rather than visual or physical ones. Responses also reveal to the user some of the internal "thinking" and "emotional state" of the voice interface.

Responses typically fall into one of three primary groups:

- **Confirmation.** The simplest form of guidance is a straightforward confirmation acknowledging understanding. In an aural environment devoid of visual cues, it's particularly important to communicate success verbally to the user.
- **Error recovery.** In a realm as error-prone as voice content, good error-recovery strategies are required for users to be able to bounce back quickly. The lack of visual cues also makes it difficult for users to know that an error has occurred unless it's explicitly stated.
- **Resource delivery.** Voice content is also a form of response and feedback for users, though it sounds very different from confirmation and error recovery. After all, it's the primary reason our users are joining us in the first place.

Confirmation

Perhaps the most common type of response issued by voice interfaces is *confirmation* immediately after the user answers a prompt with their intent. In an *explicit* confirmation, the voice interface repeats its interpretation back to the user and requests a rapid reconfirmation via a prompt:

Ask GeorgiaGov:	*Welcome to GeorgiaGov. Feel free to ask a question.*
User:	*How do I renew my driver's license?*
Ask GeorgiaGov:	***Does this sound helpful? Driver's licenses.***
User:	*Yes.*

An *implicit* confirmation, on the other hand, obliquely acknowledges the response before barreling ahead without an explicit reconfirmation from the user. Here's how the same exchange would go with implicit confirmation instead:

Ask GeorgiaGov:	*Welcome to GeorgiaGov. Feel free to ask a question.*
User:	*How do I renew my driver's license?*
Ask GeorgiaGov:	***I heard you say DRIVER'S LICENSES.*** *The Department of Driver Services helps you obtain, renew, or replace learner's permits, ID cards, and licenses. In Georgia, you must be at least sixteen years of age (with the exception of drivers with legally blind parents) to carry a full license for everyday, non-commercial vehicles.*

While implicit confirmation may seem more efficient, it may also frustrate the user, who now needs to interrupt the interface in the event of a miscommunication. For this reason, even if it makes for longer-winded voice interactions, explicit confirmation is usually the best call. Confirming user responses is crucial, because a simple misunderstanding can have a ripple effect across future points in the interface—like hearing "chemistry" instead of "Clemons Street" on an in-car system—and becomes harder to fix later on, as users burrow deeper down the rabbit hole.

Error recovery

Because *errors*, caused by a misunderstanding or a fault in logic, are common in voice interfaces, it's critical to have effective error-recovery strategies to place users back on the right track. There are several types of errors voice interface designers should be aware of:

- **No-speech timeouts** occur when the normal wait time for an expected user response ends without any speech detected.
- **Non-understanding** or **no-match errors** occur when a voice interface misunderstands a user, whether it's because the user's response is *out-of-domain* (the interface isn't designed to handle this kind of user response) or *in-domain* (by design, the interface should be handling this response appropriately but isn't doing so).

In the following example from Ask GeorgiaGov, we wrote an error-recovery strategy for whenever the user's utterance is unclear to the interface. Graceful error-recovery strategies will encourage users to come back.

Ask GeorgiaGov:	*Welcome to GeorgiaGov. Feel free to ask a question.*
User:	*(Unclear)*

Ask GeorgiaGov:	***Sorry, I didn't understand. Ask me a question related to the state government of Georgia.***
User:	*How do I renew my driver's license?*
Ask GeorgiaGov:	*Does this sound helpful? Driver's licenses.*
User:	*Yes.*

This is known as a *move-on strategy*, where the voice interface resorts to a different line of questioning entirely to solicit the same information instead of simply uttering the same prompt ("Feel free to ask a question") over and over again in a game of chicken. Here, the interface rephrases the prompt so that it specifies its desire for relevant questions ("Ask me a question related to the state government of Georgia") rather than potentially irrelevant ones ("Feel free to ask me a question," i.e., any question at all). This requires designers to ready two or more distinct dialogues to accomplish the same task, redundancy that might be warranted given its benefits for the overall user experience, as researchers Timothy Bickmore, Ha Trinh, Reza Asadi, and Stefan Olaffson argue in *Studies in Conversational UX Design* (http://bkaprt.com/vcu36/03-03).

Achieving a set of error-recovery strategies that succeeds in restoring the user's faith in the interface is essential: if users are unable to accomplish what they set out to do, as Pearl writes in *Designing Voice User Interfaces*, they likely won't ever return to your interface to give it another shot (http://bkaprt.com/vcu36/03-02).

Resource delivery

Once Ask GeorgiaGov has identified the intent of the user ("How do I renew my driver's license?"), it conducts a search and presents the first search result ("Driver's Licenses") to the user for confirmation, as we've seen before. If the user says no, it continues down the list of returned search results until the user says yes.

Now that we've identified our intent, we can perform the content delivery the user has come to us for. Ask GeorgiaGov first shares voice content drawn from the "What You Should Know" section of the page before asking whether the user wishes to go through the list of available FAQs one by one.

Ask GeorgiaGov: ***The Department of Driver Services helps you obtain, renew, or replace learner's permits, ID cards, and licenses. In Georgia, you must be at least sixteen years of age (with the exception of drivers with legally blind parents) to carry a full license for everyday, non-commercial vehicles. I have a list of related information. Would you like to hear it?***

User: *Sure.*

Ask GeorgiaGov: ***Does this sound useful? How do I transfer my out-of-state driver's license once I move to Georgia?***

User: *Yes.*

Once the list of available FAQs related to the topic is exhausted, or once the user decides they've heard enough (whichever comes first), the voice interface offers an opportunity for the user to request an agency phone number. Because only a subset of the entirety of Georgia's state government information is housed on Georgia.gov itself, the provision of an agency phone number allows residents to call hotlines directly on another device.

Ask GeorgiaGov: *Within thirty days of moving to Georgia you are required to visit the nearest Department of Driver Services Customer Services Center. Bring along your current license, proof of your Social Security number, two proofs of Georgia residency (such as a bank state-*

	ment, utility bill, or rental contract), and proof of identity (such as a birth certificate, passport, or immigration card). You'll be required to hand over your former license, pay the license fee, and receive a temporary license. You may also register to vote. Just a few weeks later, your permanent license will arrive by mail. **Would you like to hear more related information?**
User:	*No.*
Ask GeorgiaGov:	***If you need a phone number for further assistance, say PHONE NUMBER. If you have another question, feel free to ask it. Otherwise, you can say EXIT.***

As you can see, resource delivery looks very different from the normal nuts and bolts of navigating through the interface. We can tell quite clearly that the interface isn't merely reacting to something we've said; it's now sending critical information down the wire that we should pay attention to. Clearly delineating what is content and what isn't is part of the exciting challenge of crafting great dialogues.

WRITING DIALOGUE

It's impossible to overstate the importance of writing in good voice interface design. Because users will interact with your voice content with speech, your voice interface design process begins, for all intents and purposes, with a dialogue writing process. Writing dialogues also enables you to test your interactions by reading them aloud and to see how your voice content performs within them.

Sample dialogues are the key artifacts you'll need to deliver, and you'll want to have a variety of such "scripts" to work with. These "screenplays" need to reflect a representative cross section of the kinds of voice interactions you plan to facilitate. They're the intersection between the dialogue elements that undergird every voice interface—prompts, intents, and responses—and the natural narrative progression of human conversation that brings the user to their goal.

One of the challenges of dialogue writing is ensuring that the machine-friendly dialogue elements that voice interfaces need to function properly don't obscure the human-friendly narrative of our conversation—that beginning-to-end progression that concludes happily with our arrival at desired voice content.

Sample dialogues enable us to zoom in on individual utterances to pinpoint a wayward sentence or word and to zoom out to how the entire conversation will progress based on a user's series of choices. That's why it's essential to write not only successful dialogues that guide users down happy paths to their content, but also those that run into brick walls and trigger slews of errors. Sample dialogues should represent a realistic snapshot of how you think users will interact with your voice interface and your voice content—in both good and bad ways.

Writing sample dialogues doesn't necessarily mean handling every single possible permutation of how a conversation might transpire. The threshold where you have enough sample dialogues to start implementing is a cross-functional decision. For many designers, this will be a matter of honing an instinct. Just as there is a point beyond which a Photoshop comp would be better perfected as an InVision prototype or in working frontend code, there's a subtle juncture at which it makes sense to stop writing and start building.

Best practices for dialogues

When writing text for your dialogue elements, especially prompts and responses, mirror how conversation works in human environments, and author sample dialogues and interface text from the standpoint not of the writing on the page but rather of the speaking the interface will do, sounding as natural as possible in the process.

In *Designing Voice User Interfaces,* Pearl lists several best practices for voice interface design that center the organic and unpredictable nature of human speech (http://bkaprt.com/vcu36/03-02):

- **Keep it short.** Speech tends to be "ephemeral, transitory, and linear." Dialogues should avoid lengthy sentences, unfamiliar terminology, and too many options.
- **Keep it natural.** Dialogue should be as natural to the human ear as possible, using words and phrases people use when speaking, not when writing.
- **Design for ambiguity.** Voice interfaces can't simply display a list of options to disambiguate similar-sounding options. The more ambiguities are present in a dialogue, the more difficult it is to conduct disambiguation through dialogue in a natural way.
- **Support corrections.** Users should be able to repair an utterance containing a mistake by issuing a correction, even if it's easy for the user to start over from the beginning.
- **Timing is important.** Time is one of the key means by which voice interfaces can approximate physical space. Like negative space, silence is extremely effective, but pauses of longer than four hundred milliseconds sound unnatural in human conversation.

Our machines aren't naturally equipped with the sort of instinct that we humans have honed since childhood to know when to stop talking, to accept a correction, and to jump in at just the right moment. When you write sample dialogues, consider this gap between human and machine as you deploy dialogue elements—and, as we'll see next, as you humanize the interface and transform it into a natural-sounding conversationalist.

Sounding natural

Our organic, context-aware, human conversations are littered with *conversational markers,* linguistic elements that act as highway signage on the extemporaneous expressway. As transition words and rhetorical indicators that behave like physical mile markers or exit signs, they're the connective tissue that binds together disparate chunks of conversation into a cohesive end-to-end narrative.

Conversational markers are particularly useful because they can inflect responses in ways that offer rich additional insight to the user beyond the direct meaning conveyed by the response itself. Pearl identifies three types of *conversational markers* that are essential for voice interfaces because they adjust and modulate the responses you're serving your users (http://bkaprt.com/vcu36/03-02):

- temporal markers (*first* and *last*, *previous* and *next*, and *almost there* and *finally*)
- acknowledgments (*thanks* or *got it*, *understood* or *sorry*, and *didn't catch that*)
- affirmative feedback (*great job*, *nice work*, and *you did it*)

All three types of conversational markers serve to situate the user in a larger narrative of goal orientation. In many cases, the only way to test such markers is to read dialogues aloud in a live setting to isolate any foibles (like saying *last but not least* when there are still multiple steps remaining).

Voice users rely heavily on conversational markers—perhaps even more so than readers of textual content. Our real-life chats are replete with conversational markers, because they forge a human sense of empathy by adding an emotional dimension to sequential prompts, success messages, and errors. Imagine how the following conversation would sound without the emphasized words:

Doctor: *How are you feeling today?*

Patient: ***Well, first*** *I had a fever and then I started getting a cough.*

Doctor: ***I see, I'm sorry to hear that.*** *Can you breathe in slowly and hold your breath for me, for ten seconds?* ***Afterwards,*** *breathe out slowly.*

Patient: *(breathes in and holds breath)*

Doctor: **Almost there.** *Four more seconds.*

Because they instill in the user a sense that the voice interface knows how the conversation is progressing and where it's headed, conversational markers are critical to the authenticity of voice interactions. One example of this is how Ask GeorgiaGov ensures the user understands that it's repeating an agency phone number upon request. Imagine how stilted this exchange would feel without the emphasized word:

Ask GeorgiaGov: *The number for the Georgia Department of Driver Services is 404-657-9300. If you need that number again, say Repeat.*

User: *Repeat.*

Ask GeorgiaGov: ***Again****, that number is 404-657-9300. If you need that number again, say Repeat.*

Sometimes a conversational marker can reset the relationship with the user and help them feel less concerned by acknowledging a mistake:

Ask GeorgiaGov: *Welcome to GeorgiaGov. Feel free to ask a question.*

User: *(Unclear)*

Ask GeorgiaGov: **Sorry, I didn't understand.** *Ask me a question related to the state government of Georgia.*

BEYOND WRITING

As humans, we have an innate understanding of how dialogue works. Machines, on the other hand, need structures that mirror the internal workings of how they process and serve information. We now have a range of tools at our disposal—onboarding, prompts, intents, and responses—to craft compelling dialogues.

Voice interfaces juggle concepts that are wholly unfamiliar to humans engaging in normal conversation: prompts, intents, and responses that adhere to clear inputs and outputs. But it's also important to consider the human signals that decorate every organic conversation: the conversational markers that link strands of dialogue and compartmentalize conversations into comprehensible narratives.

Voice interface design, especially when it comes to voice content, is an uneasy fusion of two distinct design approaches: the *dialogues* that underpin the interface's narrative and its movement toward the right content for the user; and the *flows* along which users traverse interfaces, like rivers etching their way down watercourses. These need to happen in parallel and can't occur in isolation, as we'll see in Chapter 4.

4 DIAGRAMMING FLOWS

THE MENTAL MAPS WE BUILD and prune as we navigate interfaces inform how we find the information we seek. We deduce the right way to get to the airport security checkpoint or the right track in a bustling train station thanks to signage and signaling mechanisms in our physical world that orient and direct us. The signs, arrows, and architectural cues in transportation hubs are guideposts of *wayfinding* in the built world.

In visual experiences, orientation and navigation are couched in manual actions: clicking a mouse or hitting a key, swiping left or scrolling down. Keeping track of where you are usually comes down to some form of visual hierarchy, like sitemaps and breadcrumbs.

Wayfinding is also a key aspect of our aural and verbal landscapes, though we're concerned with time and sound rather than space. Voice interfaces are some of the more intriguing examples of user experiences navigated by sound. Rethinking physical or visual interfaces as aural ones requires us to recalibrate our entire approach to blueprinting flows that take users where they need to go.

Along with dialogue writing, carving out the flows that outline navigational and wayfinding mechanisms for users is crucial work for enabling voice content. How linear and consistent our information architectures are determines how easily users can encounter the content they're looking for in as little time and as few steps as possible.

UNDERSTANDING FLOW

In user experience, *flow* represents a user's movement through an interface—the end-to-end path they travel to achieve their mission. Though flow is a common word in user interface design, it has more nuance in voice interface design.

Flow is critical because it determines the discoverability of our voice content. The more steps required to get to a piece of content, or the more time spent getting there, the less discoverable your content becomes. We can spend all the time we like on ensuring our content is voice-ready, but if it isn't discoverable within the interface, all that work was for naught.

In other words, in voice interface design, flow dictates all possible journeys through a voice interface—especially to voice content. But to illustrate why flow is so foundational in voice interface design, let's take a look at a few examples of how well-designed flows can benefit voice experiences.

Taking inspiration from single-access keys

As a teenager in Colorado, one of the topics I enjoyed the most in biology class was forestry—not solely to admire verdant mountainsides for their evergreens and aspens, but also for the steely efficiency displayed by *single-access keys*, research tools for identifying living things that bear some resemblance to flowcharts (though with many more prescriptions).

In forestry, researchers use single-access keys for tree identification: Is this tree coniferous or broadleaf? Is it a pine or a spruce? Single-access keys closely resemble the arboreal branching structures they classify: a primary trunk divided into two or more offshoots (**FIG 4.1**).

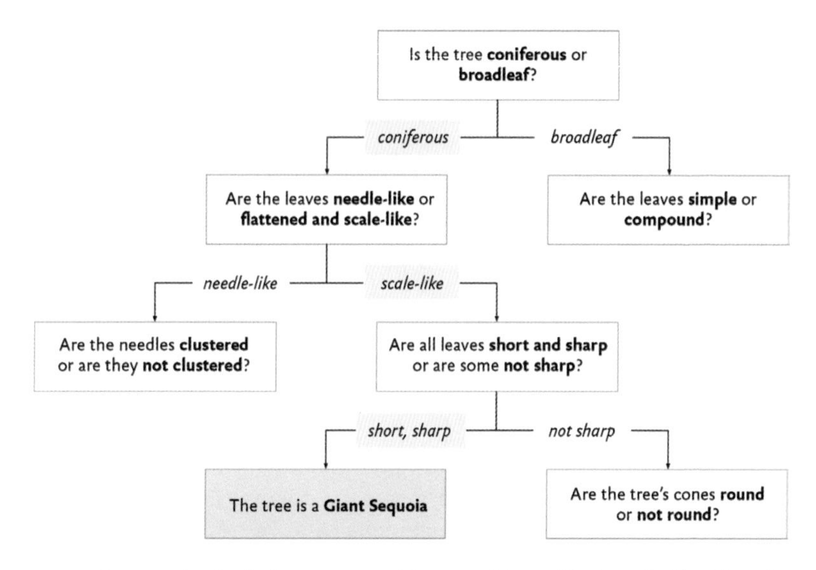

FIG 4.1: An example of a single-access key with successful identification of a Giant Sequoia, adapted from Oregon State University's "Common Trees of the Pacific Northwest" dichotomous key (http://bkaprt.com/vcu36/04-01).

The vast majority of single-access keys only have two possible responses to a given question ("coniferous" or "broadleaf," for example), preventing choice paralysis. Other single-access keys might have three or more possible options at each juncture that funnel the user ever closer to their target.

Single-access keys are instructive for voice interface designers, because when a voice interface is trying to identify a complete user intent, it plays a similar game of Twenty Questions. Moreover, single-access keys require every interaction to begin from the same single access point (hence the name!), or kickoff question: "Is the tree coniferous or broadleaf?" Voice interfaces likewise begin with the same sort of initial prompt for all users. At each step, the questions become increasingly specific to the user's previous responses.

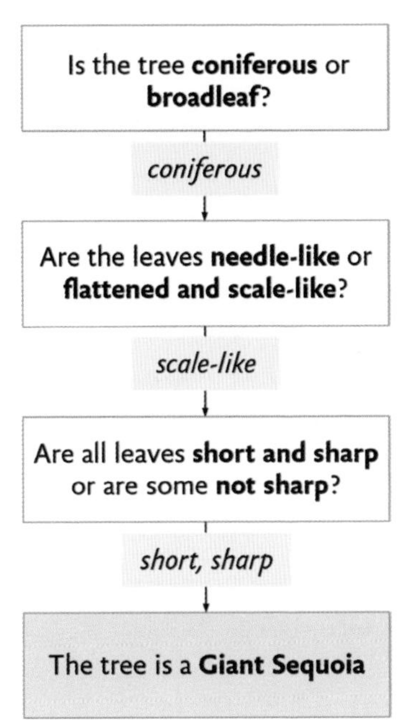

FIG 4.2: A linearized version of the same single-access key demonstrating how other areas of the "interface" are of no concern to the user—and need not be.

Single-access keys can inspire us in three crucial ways:

- First, instead of many simultaneous choices, the user only sees a few options on any given card, meaning superfluous questions from elsewhere in the decision tree never surface. In this way, single-access keys show *linearity* (**FIG 4.2**). Each step deeper into the key doesn't increase structural complexity; rather, it carefully discloses greater specificity in a one-by-one cadence.
- Second, single-access keys also display considerable *unidirectionality*. Tree taxonomists can only use a one-way flow; it becomes nonsensical in the reverse direction. A sort of gravitational pull propels the user to a successful identification of a tree's genus. But "unidirectionality" doesn't mean we've lost the ability to jump back up and correct course when stuck. If you answered a question wrong, all you have to do is step back and respond differently (**FIG 4.3**).

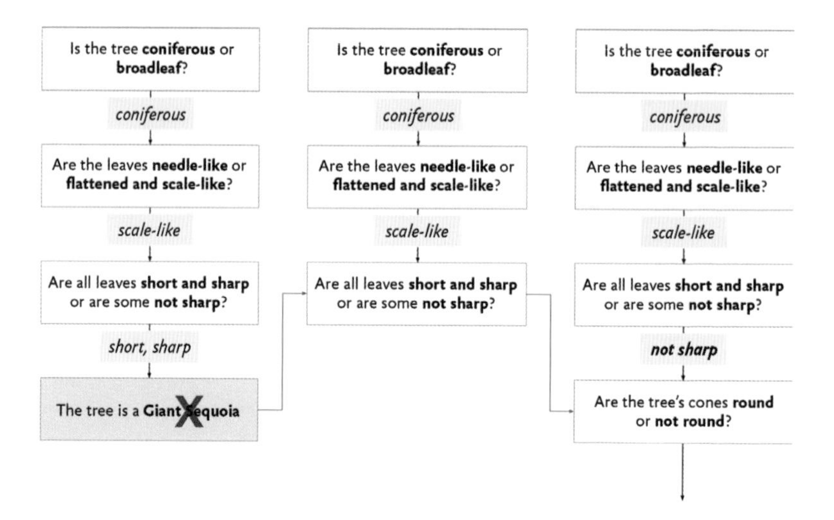

FIG 4.3: Single-access keys are also uniquely well-suited to course correction: users can return to an earlier point in the interaction to correct earlier responses.

- Finally, their *consistency* makes single-access keys eminently learnable. Each answer notches a successful outcome in the user's mind, thus refining the user's mental model of the interface and likely accelerating subsequent interactions.

The linear flow, the gravity-like unidirectionality, and the cross-interaction consistency of single-access keys all apply to voice content. The best voice interfaces are those that help users spelunk ever deeper into the interface, with a life-saving rappelling rope to return to previous forks in the cave.

Lines, not networks

In websites, sitemaps and navbars help users quickly scan lists of links and venture in and out of smaller branches, but they also represent a road atlas cataloguing their full geography—a *hub-and-spoke* model, generally with the homepage functioning as the central landing point. In voice interfaces, however, users might build only a tenuous mental map of the interface and,

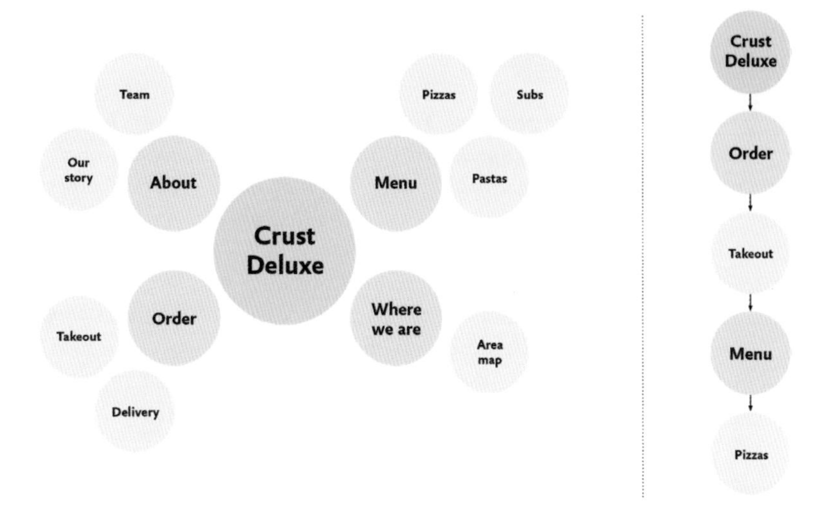

FIG 4.4: Information architectures for a hypothetical website and voice interface for Crust Deluxe. At left, the website's navigation structure mirrors a hub-and-spoke network typical of sitemaps. At right, the voice interface's navigation structure matches the one-way linearity users expect from voice interfaces.

at worst, may find the entire thing unchartable. We need to translate the hub-and-spoke sitemap into a more *linear* voice-friendly decision tree.

Achieving linearity in voice interfaces means moving away from the navigational structure suggested by a hierarchical sitemap, with its hubs and spokes, and instead situating the key places scattered throughout your interface within a unidirectional flow.

For a pizza restaurant, for example, a website might allow a user to choose between browsing the menu and placing an order. Its voice interface, however, might prompt the user to express their intent to place an order before reciting the restaurant's menu at all, for efficiency's sake (**FIG 4.4**).

This sort of emphasis on unidirectionality enables users to think of your voice interface as a template for a conversation rather than as an analogue to traditional graphical user interfaces. The less we experience an overt hierarchy, the better our voice usability, as chatbot designer Eunji "Jinny" Seo writes in *Chatbots Magazine*:

> *Prune your branches. For conversational interfaces, linear navigation is superior to nested navigation...conversational interfaces should not be direct translations of graphical user interfaces. (http://bkaprt.com/vcu36/04-02)*

A linear information architecture engaging in progressive disclosure also means that it's much easier to facilitate *happy paths* for users with a clear purpose or destination in mind. "Often times [*sic*], it's better to put the users onto a happy path and offer ways [to] veer off it, than to show the users *all* the functionalities up front," continues Seo (her italics). This is why, in *Don't Make Me Think*, usability researcher Steve Krug characterizes conversational navigation as a *funnel* that presents the most necessary actions first, before yielding more specialized ones (http://bkaprt.com/vcu36/04-03).

Taking our visual hierarchies and translating them into linear flows involves both crafting dialogues and outlining flow. In the process of writing dialogues, designers should consider how to linearize what is normally rendered hierarchically on websites. For Ask GeorgiaGov, for instance, we decided to take each nested level of information and put those steps in a sequence; users go deeper by traveling farther along a single line, not selecting from a complex tree.

Now, we need to visualize it in a way our engineers or development platform will understand. How do we as voice interface designers actually *diagram* flow so it's comprehensible to both stakeholders and voice interfaces? To answer this question, we have to go back in time to the earliest examples of flow in interactive voice response systems.

CALL-FLOW DIAGRAMS

The term *call flow* originates from IVR systems that acted as virtual switchboards to route users to the correct extension or section of an interface. *Call-flow diagrams* are schematic blueprints that depict every intersection, every branch, and every path that the user can take within a voice interface, much like plumbing layouts or circuit maps.

A call-flow diagram is a "layout of conceptual relationships," Harris writes in *Voice Interaction Design*, not a "coding blueprint" or an engineering specification (http://bkaprt.com/vcu36/02-02). For us as designers, call-flow diagrams help us visualize the overarching structure and available paths through our voice content. For the engineers and frameworks we work with, call-flow diagrams are the maps that synthesize our dialogue elements and voice content into a roughly machine-ready structure. In other words, sample dialogues and call-flow diagrams together represent the final step before a handoff to a developer team, after which they're translated into machine code.

But what's the relationship between the sample dialogues we just wrote and call-flow diagrams? Sample dialogues represent journeys in voice interfaces, but designers seldom have a sample dialogue at the ready for every conceivable flow. Instead, voice interface designers refer to multiple sample dialogues to outline as many journeys as possible within a single call-flow diagram. One sample dialogue is seldom enough to account for every possible user journey in the schematic.

Call-flow diagrams are just as essential as sample dialogues in the voice interface design process, and they are also deeply intertwined. For this reason, articulating call-flow diagrams early in the design process supports designers who prefer to iterate on both sample dialogues and call-flow diagrams at the same time. As Harris writes:

> *Traditional development cycles have tended to put the words last, in a rushed and unguided way, as internal deadlines and release dates loom....On the other hand, nor can the words come first....[W]ithout the plot component—the call flow—the dialogue can only go so far.* (http://bkaprt.com/vcu36/02-02)

As such, writing sample dialogues and building call flows must occur in parallel and intersecting ways that consider their mutual impacts on each other.

Call-flow diagrams are also valuable in helping determine how discoverable a piece of voice content is. The deeper a content item is in a flow diagram, the more steps are required to get to it, and therefore the less discoverable it is.

There is no single right way to design a call-flow diagram. Literature on voice interface design recommends a diverse range of diagramming approaches, and there is great variation between accepted patterns for call-flow diagrams depicting chatbots, voice assistants, and IVR systems. Nonetheless, certain unifying threads extend from the early days of call-flow diagramming in IVR systems to the modern era of voice assistants, especially the basic elements: the *nodes* and *arrows* that represent flow, and the *states* that represent each node.

Nodes and arrows

Call-flow diagrams use two visual elements to establish flow: *nodes* that represent each stopping point and each juncture in the voice interface, and *arrows* that represent moves between those junctures.

In call-flow diagrams, nodes are typically drawn as cards. Each node is an important intersection in the interface. You can think of nodes as "stoplights" or "crossroads" where some input is required from the user or where some action is required from the interface.

Nodes are convenient ways to visualize depth and duration, two of the traits that characterize discoverable voice content, because each node represents another step toward arriving at an item of voice content—and each additional node means a greater depth and a longer duration.

Nodes are connected by arrows, which reflect motions from one node to another (**FIG. 4.5**). Arrows represent transition on the part of the user or the voice interface from one place to another in the interface.

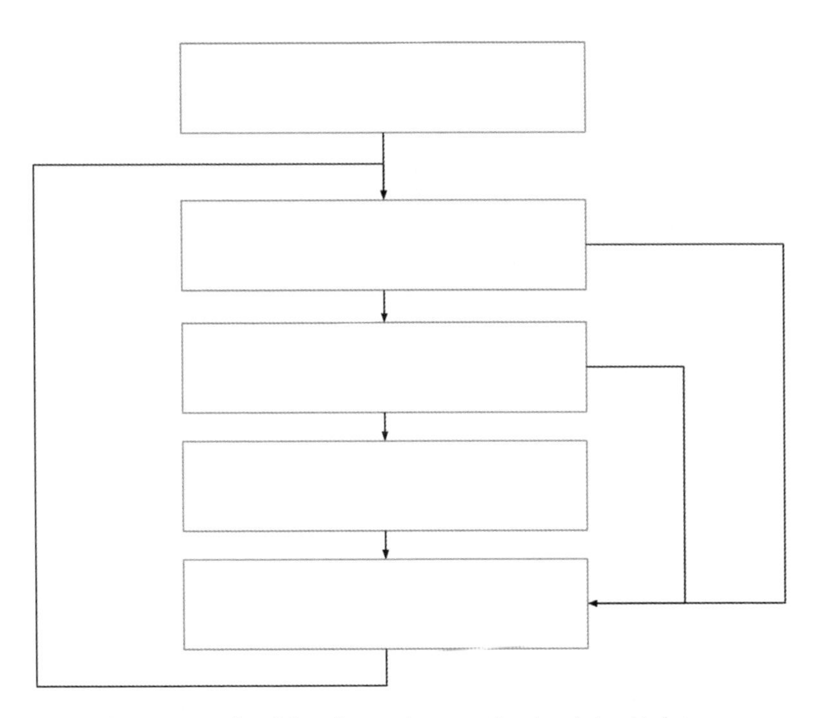

FIG 4.5: An empty sample call-flow diagram demonstrating the relationship between nodes and arrows.

Thus, each arrow represents a change in context—a shift in how the user or interface interacts with its counterpart. Nodes and arrows are useful for drawing flow diagrams, but they're also abstract concepts that we need to map gracefully onto our sample dialogues.

Depending on what type of call-flow diagram you're creating, nodes express either *decision states* that map every interface action, or *dialogue states* that chart the user's intent.

Ask GeorgiaGov:	Welcome to GeorgiaGov. Feel free to ask a question.	**Prompt for question** Decision state
User:	How do I renew my driver's license?	
Ask GeorgiaGov:	*[forwards topic to search, invisible to user]*	**Forward topic to search** Decision state
Ask GeorgiaGov:	Does this sound helpful? Driver's licenses.	**Prompt if topic is correct** Decision state
User:	Yes.	
Ask GeorgiaGov:	Within 30 days of moving to Georgia you are required ...	**Serve topic content** Decision state

FIG 4.6: The first few lines of our sample dialogue rendered as decision states. Each decision state represents an action that the interface undertakes.

Decision states

I define *decision states* as nodes that represent each individual action by the interface. If the interface prompts the user, that's a decision state; if the interface needs to conduct a search based on a user's response, that's also a decision state. For this reason, your call-flow diagram may sometimes need to account for machine actions that aren't represented in your sample dialogues or otherwise aren't visible to the user.

Here's an example of portions of a sample dialogue marked as decision states (**FIG 4.6**). Remember that what we're concerned with here is interface actions. To make each decision state node easier to read, especially if some of your prompts and responses are on the lengthy side, I recommend describing the decision state in a few short words. This will make the eventual call-flow diagram much more legible.

Decision states show a high degree of granularity. Because they care about each individual utterance the interface issues or action it executes, call-flow diagrams showing decision states help designers zoom in on elements like error-recovery strategies and other small-scale feedback that might be glossed over in a more sweeping call-flow diagram.

Ask GeorgiaGov:	Welcome to GeorgiaGov. Feel free to ask a question.	
User:	How do I renew my driver's license?	
Ask GeorgiaGov:	*[forwards topic to search, invisible to user]*	**Identify desired search topic** Dialogue state
Ask GeorgiaGov:	Does this sound helpful? Driver's licenses.	
User:	Yes.	
Ask GeorgiaGov:	Within 30 days of moving to Georgia you are required ...	

Ask GeorgiaGov:	I have a list of related information. Would you like to hear it?	**Identify desired related content** Dialogue state
User:	No.	

Ask GeorgiaGov:	If you need a phone number for further assistance, say *Phone number*. If you have another question, feel free to ask it. Otherwise, you can say *Exit*.	**Identify if phone number is wanted** Dialogue state
User:	Yes, phone number.	

FIG 4.7: Our sample dialogue rendered as dialogue states. Each dialogue state represents an intent-identification process and realization of that intent.

Dialogue states

On the other hand, some designers and development frameworks prefer a larger-scale call-flow diagram. *Dialogue states* are nodes that each represent a complete intent-identification process by the voice interface and the resolution of that user's intent. All decision states for that intent—all error-recovery strategies, user corrections, and machine actions toward resolution—are subsumed into a single dialogue state.

In *Voice User Interface Design*, Cohen, Giangola, and Balogh define dialogue states as "a single interchange between the caller and the system" that hinges on a user's overarching intent, like conducting a search or requesting a phone number (http://bkaprt.com/vcu36/01-02). In this case, the entire exchange involved in conducting that search or requesting that phone number represents a dialogue state, including all false starts by the user and all error-recovery strategies to correct that input.

To show this in action, here's an example of a sample dialogue rendered into dialogue states (**FIG 4.7**). As you can see, each dialogue state represents the complete exchange required to identify an intent and deliver an appropriate response, with each individual decision state subsumed into the larger dialogue state. In short, much of the detail is swept under the rug in favor of understanding the larger sections of the interface that each realize a single user intent, like searching for a topic or retrieving a phone number rather than recovering from an error or conducting a search.

While dialogue states help designers conceptualize each general task in the voice interface, they display very low granularity. Generally, dialogue-flow diagrams don't have the capacity to illustrate complex error-recovery approaches like the move-on strategy. Instead, they allow designers to see an overview of the "regions" or "sectors" of their interface that handle different concerns, rather than the blow-by-blow account depicted by decision-flow diagrams.

Decision-flow diagrams versus dialogue-flow diagrams

While *decision-flow diagrams* spotlight the relationships between interface actions (decision states), *dialogue-flow diagrams* illuminate the relationships between user intents (dialogue states) and are made up of one or more decision states.

For voice interfaces that require lots of small decisions here and there, decision-flow diagrams can shed light on the low-level design issues. And because decision-flow diagrams are meant to represent the full slate of possible journeys, including error states, it's much easier to translate sample dialogues directly into them. They're also helpful for low-level reviews where scrutinizing every action is critical—for compliance reasons, for instance.

Meanwhile, dialogue-flow diagrams tell the overarching story—how the user might explain the interface to a friend. They are optimal for interfaces that deal with a large array of intents, and can be useful for understanding how users switch gears between different intents, whose resolution might end with an open-ended prompt like: "What do you want to do next?"

Different implementation tools demand different diagramming approaches. Because of this, you'll want to consult your engineers and architects to learn what the development tool of choice expects: Is a dialogue-flow diagram, a decision-flow diagram, or some mix between the two most appropriate?

Some IVR frameworks and omnichannel development platforms, like Dialogflow, provide graphical user interfaces (GUIs) for hybrid flow diagrams that mix and match decision states and dialogue states, while others take the approach of IBM's Watson Assistant and focus on dialogue states, offering canned error-recovery strategies with limited configuration of internal decision states. Other tools, like chatbot frameworks, operate solely in decision states. Still others, especially developer-oriented frameworks like the Alexa Skills Kit and Oracle Digital Assistant, don't provide any means to diagram call flows and instead require all flows to be translated into lines of code or form fills.

If your choice of technology is still in flux, my recommendation is to start with a decision-flow diagram that accounts for all of the prompts and responses first before moving on to a dialogue-flow diagram. This is because many of the smaller elements—prompts and responses in decision states—will inform how the broader intent-identification processes unfold within dialogue states. It's best to have the details set in stone before sketching the bigger picture.

How sample dialogues and call-flow diagrams are translated into interface code is a matter of negotiation between you and your engineering team. Depending on the development approach, call-flow diagrams can be useful for both designers and engineers—but for us as designers, they allow us to visualize and iterate on voice interactions in compelling ways.

DIAGRAMMING FLOWS

To draw an initial draft of a call-flow diagram, let's take a look at the first few lines of the sample dialogue we ended with in Chapter 3 for Ask GeorgiaGov. Because the Alexa Skills Kit

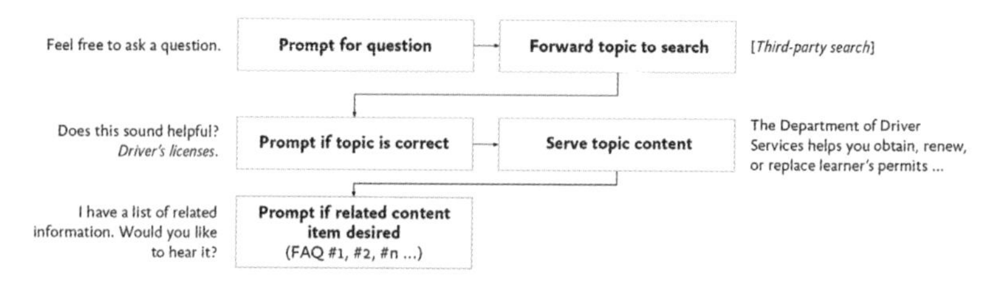

FIG 4.8: The first few lines of our sample dialogue rendered into a call-flow diagram. Note that we've placed prompts on the left and responses on the right.

doesn't have a GUI, we opted to focus first and foremost on a call-flow diagram handling decision states. In this sample dialogue, we have an initial machine prompt, a third-party CMS search returning a response, a confirmation prompt, and finally resource delivery (a form of response).

Ask GeorgiaGov: *Welcome to GeorgiaGov. Feel free to ask a question.*

User: *How do I renew my driver's license?*

Ask GeorgiaGov: *Does this sound helpful? Driver's licenses.*

User: *Yes.*

Ask GeorgiaGov: *The Department of Driver Services helps you obtain, renew, or replace learner's permits, ID cards, and licenses. In Georgia, you must be at least sixteen years of age (with the exception of drivers with legally blind parents) to carry a full license for everyday, non-commercial vehicles. I have a list of related information. Would you like to hear it?*

Since all user responses will be user-generated and therefore unpredictable, let's not worry about those for now. I recommend that newcomers to call-flow diagrams put prompts on the left and responses on the right to make things easier to visualize:

PROMPTS	RESPONSES
"Welcome to GeorgiaGov. Feel free to ask a question."	"Driver's licenses." (returned from third-party search)
"Does this sound helpful? Driver's licenses."	"The Department of Driver Services helps you obtain, renew, or replace learner's permits..."
"I have a list of related questions. Would you like to hear it?"	

Though some designers prefer to write the snippet of dialogue itself into the card representing a decision state, this can get out of hand fast if your prompts or responses are on the lengthy side. I prefer to summarize what the machine action is accomplishing in a few short words ("Does this sound helpful? Driver's licenses" becomes "Prompt if topic is correct"). For our purposes, we can identify what each of these prompts and responses is doing and make that the description of the node.

To keep a reference in the margin, we can also use real examples of authentic dialogue from the voice interface to make it easier to navigate (**FIG 4.8**). This is something I do in all of my own call-flow diagrams to have an easy catalog to consult.

Recursive behaviors

The next part of our sample dialogue deals with *recursive behaviors* (defined as loops that end once the machine and user come to an understanding) that revolve around individual machine and user decisions. For instance, after the initial search, if a user responds "No" to a confirmation that a search result is correct, the interface moves on to the next one. Consider a scenario in which the first topic the interface offers up isn't what the user wants.

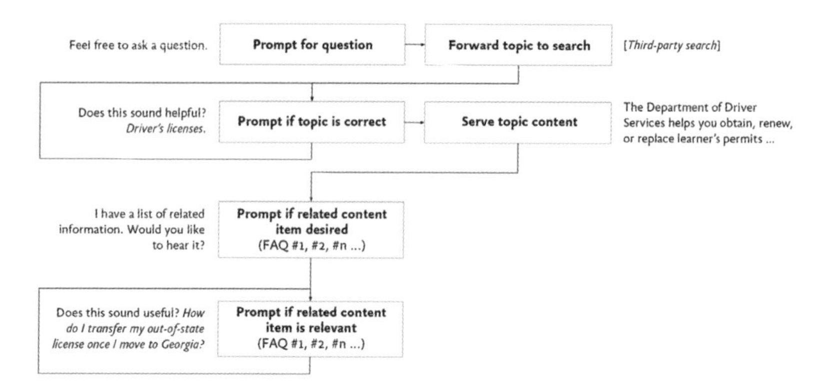

FIG 4.9: To illustrate recursive decision states that loop until the user decides to stop, we can use arrows to depict loops. Notice the two loops for "Prompt if topic is correct" and "Prompt if related content item is relevant."

Ask GeorgiaGov:	*Welcome to GeorgiaGov. Feel free to ask a question.*
User:	*How do I renew my license?*
Ask GeorgiaGov:	*Does this sound helpful?* **Getting a boating, fishing, or hunting license.**
User:	*No.*
Ask GeorgiaGov:	*Does this sound helpful?* **Driver's licenses.**
User:	*Yes.*

Similarly, when the user reaches the FAQs section of the interface, if the user determines a question to be irrelevant, the interface moves on to the next question. These sorts of "loops" are easily accounted for in call-flow diagrams illustrating decision states.

Ask GeorgiaGov: *Does this sound useful?* **How do I transfer my out-of-state driver's license once I move to Georgia?**

User: *No.*

Ask GeorgiaGov: *Does this sound useful?* **How do I change my address or name on my driver's license?**

User: *Yes.*

Because they illustrate recursive behaviors, we can diagram these sorts of decision states as loops that continue to return the user to the same decision state until the user confirms the interface can proceed (**FIG 4.9**).

Resource delivery

Finally, we need to diagram decision states for the *resource delivery* of each FAQ answer and the phone number prompt at the end of the sample dialogue, along with the repeat-if-needed step.

Ask GeorgiaGov: *Within thirty days of moving to Georgia you are required to visit the nearest Department of Driver Services Customer Services Center. Bring along your current license, proof of your Social Security number, two proofs of Georgia residency (such as a bank statement, utility bill, or rental contract), and proof of identity (such as a birth certificate, passport, or immigration card). You'll be required to hand over your former license, pay the license fee, and receive a temporary license. You may also register to vote. Just a few weeks later, your permanent license will arrive by mail. Would you like to hear more related information?*

User:	*No.*
Ask GeorgiaGov:	*If you need a phone number for further assistance, say Phone number. If you have another question, feel free to ask it. Otherwise, you can say Exit.*
User:	*Yes, can you give me a phone number?*
Ask GeorgiaGov:	*The number for the Georgia Department of Driver Services is 404-657-9300. If you need that number again, say Repeat.*
User:	*Repeat.*
Ask GeorgiaGov:	*Again, that number is 404-657-9300. If you need that number again, say Repeat.*
User:	*Exit.*

Resource delivery might also include recursive behaviors like pagination, particularly for long tracts of content that need to be broken up into more voice-legible chunks. These should also be reflected as decision states in our call-flow diagram.

Finally, we need to account for all of the *exit strategies*: how can users return to the initial prompt of the interface, or exit the interface altogether? We can depict exits as arrows that point the user back to the onboarding step and initial prompt, which can optionally occur either at the phone number prompt ("Otherwise, you can say *Exit*") or at the repeat-phone-number prompt ("If you need that number again, say *Repeat*").

Our final call-flow diagram shows the full interaction, with all of our recursive decision states and exit strategies accounted for (**FIG 4.10**). And there you have it!

Our call-flow diagram is looking pretty good, and it helps us make granular and focused choices about how individual decision states will lead users in the right direction to their voice content. To make the call-flow diagram more legible, I generally remove the snippets of dialogue represented as marginalia once they're no longer needed.

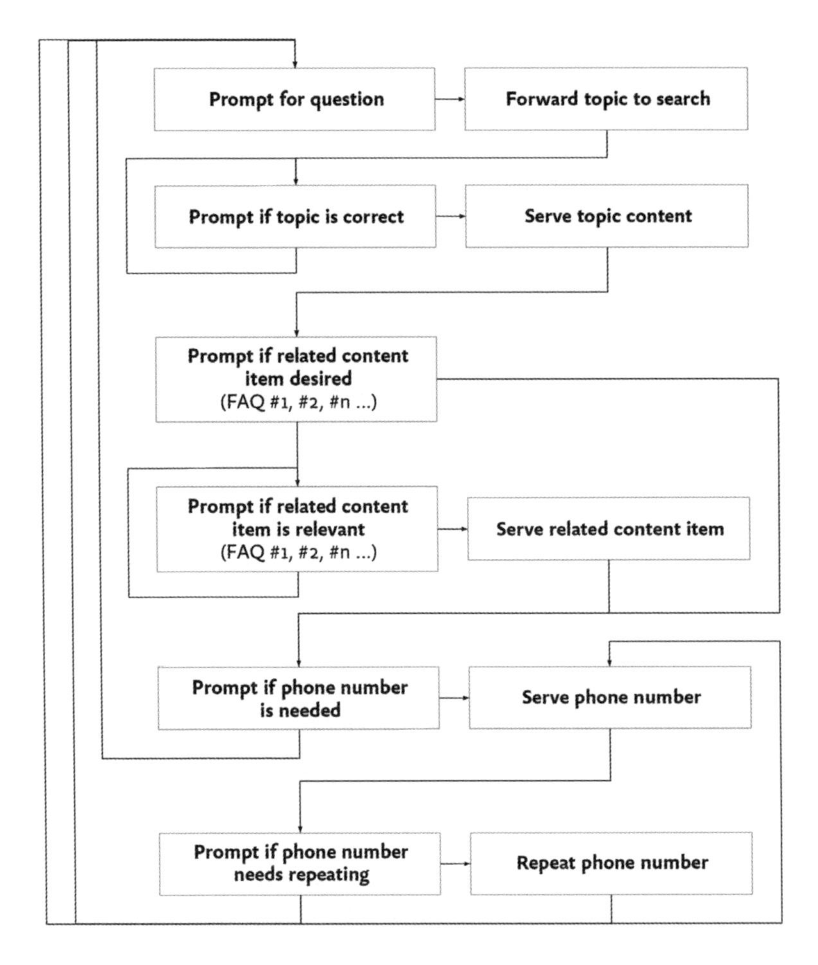

FIG 4.10: Our completed call-flow diagram. Note the additions of the phone number loop and the exit strategies to return to the onboarding and initial prompt.

This call-flow diagram doesn't dispense with any of the details. It tells us everything we need to know about each atomic unit in the call flow and helps us zoom into minuscule design considerations. Decision-flow diagrams support a deep dive into the call flow of your interface so you can see the nit-

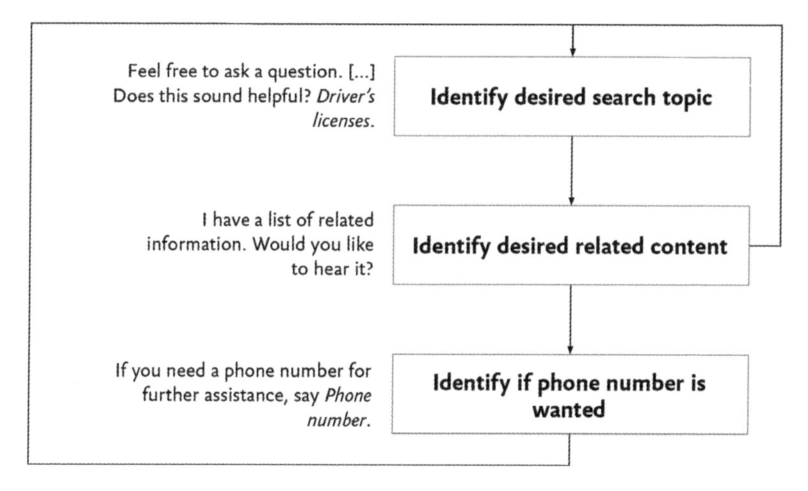

FIG 4.11: The dialogue-flow diagram we completed as an exercise for Ask GeorgiaGov. Each dialogue snippet here represents the initial prompt that kicks off the intent-identification process for that intent.

ty-gritty details in all their glory: the individual prompts and responses, recursive decision states, and exit strategies.

In some cases, however, this kind of diagram might verge on too much detail and become overwhelming. Moreover, while it tells us how users will go from one prompt to another, it doesn't demonstrate anything about how users will move between different intents. A dialogue-flow diagram, like the one I created as a thought exercise for Ask GeorgiaGov (**FIG 4.11**), offers a more zoomed-out version of your call-flow diagram and might be more appropriate for your engineering team, depending on what you can control.

VOICE NAVIGATION IS ESSENTIAL BUT HARD

Navigation and wayfinding are critical for those who use aural and verbal cues to move through interfaces. Today, as more users begin to learn voice interfaces and become as accustomed to navigating by voice as they do with a mouse, robust aural wayfinding will become even more essential.

As we've continued to narrow the gap between design processes and the eventual build itself, we've seen how architectural and engineering decisions increasingly infiltrate our thinking in terms of what to deliver as designers. Different teams and tools—and different skill sets and software choices—demand different approaches.

There's no one hard-and-fast rule about using decision-flow diagrams over dialogue-flow diagrams, nor any prohibition against fusing them together and mixing them where needed. The only prescription universally shared by voice interface designers is to advance your call-flow diagrams incrementally and iteratively alongside your sample dialogues to optimize for both rhetoric and structure simultaneously. Success in both of these deeply intertwined areas will benefit your voice content immensely during usability testing and launch preparations, the subjects we turn to in the next chapter.

5 READYING VOICE CONTENT FOR LAUNCH

THE VERY NOTION OF making a voice interface *usable* is complicated by the fact that voice differs so considerably from other types of user interfaces. After all, we need to cater not only to disabled users treating our interfaces as an alternative means of access, but also to users who may never have encountered a voice interface before.

Making our voice content usable starts with understanding the target demographic and how they currently retrieve their information, and it continues with designing an interface faithful to the expectations of both novice and experienced users. As such, some form of usability assessment needs to be built into every step of the process.

In this chapter, we'll cover voice usability and the final stages before launch: the critical phase of voice usability testing—a far cry from usability testing in other contexts—as well as the prerelease phase that prepares our interface for prime time. We'll also look at how to make sure our content is in all the right places and how to set up logs and analytics. Finally, we'll explore ways to keep improving on usability by watching how users interact with our content in the real world.

VOICE USABILITY

The people using our voice interfaces are more likely to succeed if we focus on the ways they will interact with the content within from a variety of lived experiences. Without deeply understanding our users and catering clearly to their requirements, our voice content won't quite fulfill their needs. Early, iterative prototyping will quickly pinpoint any weak spots in the interface where users may struggle.

Building usability in by default

Build usability into every point of the project timeline, including the earliest discovery phases. This could mean conducting background research on the desired audience of your interface and learning about their habits and ideals, even before embarking on articulating a content strategy, conducting a content audit, and writing sample dialogues.

If your voice interface is designed to reduce the burden on customer service agents handling dozens of conversations a day, simply sitting in on a typical shift to listen in on calls can reveal how real users will eventually interact with a voice interface. How do customers negotiate with call handlers? How quickly do they expect a solution to their problem? What are the sorts of questions they ask, and how do they word those questions?

In *Designing Voice User Interfaces,* Pearl also recommends this approach (http://bkaprt.com/vcu36/03-02). She writes that frontline customer service workers "can provide a huge amount of valuable information" because they "know about what users are really calling in about, what the biggest complaints are, and what information can be difficult for callers to find." Without real-world input, it becomes harder to cater to your users' requirements.

If early research conducted with real-world conversations yields better dialogues in the long run, so does early prototyping with authentic sample dialogues. This often helps designers fine-tune the personality of the voice interface according to user expectations.

Personality is an important consideration as well. According to voice interface designer Susan Hura, we humans tend to see voice interfaces without human personalities as "robotic or incoherent." The interface should take on the role of a conversation partner and be capable of authentic dialogues with users as early into the implementation as possible; if the interface's personality seems abrupt or otherwise uncannily not quite human, set aside sufficient time to address that back in the dialogue scripts (http://bkaprt.com/vcu36/01-03, PDF).

Too often, teams err by leaving usability research for the very end of a project, when it's far too late to implement any findings, or by conducting too little discovery at the outset. Whether you study actual conversations between humans or iteratively prototype with sample dialogues and call flows, building these processes into your earliest design and implementation phases can help you avoid returning to the drawing board later on.

Voice accessibility and usability go hand in hand

When we talk about the usability of voice interfaces, we can't ignore some of their most experienced users: people who use screen readers and other text-to-speech devices. Accessibility and usability in voice interfaces go hand in hand, and some people, like Deaf or Deaf-Blind users who require multimodal solutions with a visual or tactile component, can't use voice-only interfaces at all.

Accessibility has long been a bulwark of web usability, but it hasn't received quite as much attention when it comes to voice usability. And while the voice interface may seem an anomaly among devices traditionally thought of as assistive, accessibility advocates nonetheless acknowledge their potential. Digital Services Georgia, for instance, cited accessibility as a major motivator for making their content available through voice in addition to the web. It was a logical stepping-stone in their continuing journey to serve every Georgian.

Because screen readers recite every single piece of text and media on a page, they often overshare by including superfluous information (think "Skip to main content" or overly wordy image descriptions), lengthening the user's interaction. And because screen readers on the web are rooted in visual, page-based structures, voice interfaces can help users target a single paragraph rather than forcing them to sift through a long page of irrelevant information.

Beyond screen readers, however, designer Bo Campbell has pointed out that voice interfaces can serve distinct access needs along a number of other fronts. People with dyslexia, for instance, often find voice interfaces to be more pleasant than dealing with written text. For teams with control over speech synthesizers, volume or speech rate controls can aid hard-of-hearing users. Meanwhile, for those able to adjust how speech is recognized, a voice interface's ability to interpret slurred, shaky, or broken speech by people with cognitive disabilities can enrich their user experience. And even though drilling into the inner workings of voice assistants is seldom possible, focusing on a "linear, time-efficient architecture" can help users with cognitive disabilities understand context and wayfinding with minimal friction (http://bkaprt.com/vcu36/05-01).

While the potential for accessibility with voice interfaces remains mostly unexplored, I look forward to a future, closer than we might think, where voice interactions become full-fledged conduits through which disabled people can access content they need. The spread of accessibility into other user experiences will also heighten the demand for *multimodal accessibility,* which must reach well beyond trappings of the accessible web like the screen reader.

VOICE USABILITY TESTING

Because voice interfaces are the most humanlike interfaces, many of the traditional strategies for usability testing go out the window. Most usability tests involve some form of real-time conversation during the interaction, which might lead to corrupted data in voice usability tests. Because methodologies for voice usability testing remain new, it's important to mitigate risk by preparing diligently for their uniqueness.

Running a voice usability test also presents challenges when it comes to appropriate procedure. Test environments, test subject demographics, study questionnaires, and scenario definition all require paradigms that revolve around voice interactions. Voice usability tests require two moderators—one to watch and ask questions, the other to transcribe and catch signs of trouble. It's also best to have a developer in the room, not only to find areas for technical improvement but also to debug or troubleshoot any problems arising during the test.

Nevertheless, lessons from other kinds of usability tests still apply. The mantra "test early, test often" applies to interfaces of any type. The objective should always be to build an interface where your users don't mind spending time—a *habitable* user experience.

Early usability testing

If your voice interface isn't yet working, or if there's still significant work left to be done, usability testing can still happen outside of a formal test environment. Some approaches to this include:

- **Table reads.** Often, certain issues spring up only once users begin to interact with a voice experience in unexpected ways. *Table reads* of sample dialogues can isolate such issues. "You might have some items in your design (such as handling pronouns or referring to a user's previous behavior) that will take more complex development," writes Pearl in *Designing Voice User Interfaces*, "and it's important to get buy-in from the outset and not surprise [stakeholders] late in the game (http://bkaprt.com/vcu36/03-02)."
- **Wizard of Oz testing.** In *Wizard of Oz (WOz) testing*, the interface isn't yet ready for a full regimen of testing, so a human stands in "behind the curtain" to present the illusion of a fully functional system. The "Wizard" needs to be a researcher who deeply understands the sample dialogues and can react rapidly to a user's action with an appropriate response. One of the key benefits of WOz testing is the

ability to test early and often without all components of the interface in place.

- **Hallway testing.** Once you have a smoke-and-mirrors prototype in hand or a working build, you can conduct *hallway testing,* which involves mustering work colleagues, family members, friends, or neighbors to participate in an informal study. Because it's rare that your recruits will be members of the target demographic, hallway testing can provide early insight, but like table reads and WOz testing, it pales in comparison to the real thing.

If your voice interface isn't yet done or still needs more development to be ready for a full round of usability testing, these techniques can help you gain insights before the real deal.

Test environments and subjects

Because the primary medium of voice interfaces is sound, one risk usability researchers new to voice have seldom had to consider is aural interference. Poor soundproofing or outside noise can halt a voice usability test before it even begins. For this reason, I recommend holding tests in recording studios or similar soundproof spaces to ensure as little outside noise as possible. Though absolute silence is ideal, a mostly silent room is often the best we can do.

Preparing subjects who might be used to other types of usability tests can be challenging. Lengthy silences are common during setup, debugging, and transitions between tasks, and may feel awkward to participants—especially if they're not expecting them. For Ask GeorgiaGov, we asked participants to stay silent when not speaking with Alexa to avoid overfilling our audio recordings and transcripts. Users and moderators who are new to voice usability tests will need onboarding and training about how they need to refrain from being as chatty as they might be in other types of evaluations, and how voice interface errors surface differently from visual alerts.

When testing Ask GeorgiaGov, we found that most of our tech-savvy coworkers had little to no exposure to voice assistants in the first year or two after Amazon Alexa's launch, when it was still a novelty. Ultimately, this inexperience helped us achieve results that came close to how we envisioned actual Georgians interacting with the interface. Nevertheless, if we could repeat our project, we'd try to recruit real users in Georgia across a variety of demographics.

Trust and consent are key in any researcher-subject setting, and if your content deals with potentially traumatic topics, you must shield your users by ensuring they're comfortable with performing those tasks. State governments routinely grapple with resident inquiries dealing with abuse and trauma in areas like sexual harassment, firearms, and the carceral state. To avoid causing inadvertent harm in our Ask GeorgiaGov testing, we prescreened users with a consent form so anyone could opt out of tasks they might find distressing.

Consider the ways a task prompt might inadvertently surface deeply personal trauma: researching how to get a firearm, searching for "prison visitation," or studying child support laws won't be neutral requests for people who have been impacted by gun violence, racist incarceration practices, or domestic upheaval. By clearly distinguishing between benign and potentially harmful tasks, we were able to keep our test subjects safe.

Scenario and task definition

At the start of each evaluation, test administrators should share a *scenario* (a situation that needs resolution) and *task* (an action that needs performing) with the test subject. Many voice usability researchers define scenarios and tasks in such a way that subjects have relative free rein over their interactions, with any necessary guardrails defined.

Note how these task definitions from Ask GeorgiaGov never reveal the controls or navigation required to access the requested content. Instead, the task invites users to explore the interface and orient their mental maps on their own.

> *You have a business license in Georgia, but you're not sure if you have to register on an annual basis. Talk with Alexa to find out the information you need. At the end, ask for a phone number for more information.*

You've just moved to Georgia and you know you need to transfer your driver's license, but you're not sure what to do. Talk with Alexa to find out the information you need. At the end, ask for a phone number for more information.

If a participant faces multiple scenarios to resolve, write Cohen, Giangola, and Balogh in *Voice User Interface Design*, randomizing their sequence for each new subject is also crucial: earlier tasks can influence how the user reacts to later ones, potentially skewing the results (http://bkaprt.com/vcu36/01-02).

Retrospective testing

Usability.gov delineates two categories of testing: *concurrent*, where the researcher gathers data live during an interaction; and *retrospective*, where the researcher saves the data gathering for after the interaction is over (http://bkaprt.com/vcu36/05-02).

Many usability researchers use concurrent testing techniques that track the user's reactions and interactions in real time, but these tactics are less useful when testing voice interfaces. Spoken words relevant to the evaluator but not to the interaction at hand can accidentally invoke wake words (Alexa, for instance, could mishear "I selected" as "Hi Alexa"). This leaves the user juggling two conversations: one with the voice interface, and another with the human evaluator. A retrospective technique allows these conversations to occur at separate points in time, vastly simplifying the experience on both sides of the table.

Retrospective approaches do have frustrating flaws. For instance, humans are notorious for having slippery memories. Test subjects may insert false recollections into their impressions or misinterpret the outcome of their conversations. But there are advantages, too: retrospective approaches give users time to ponder and polish their thoughts rather than doling out tidbits in an incremental stream of consciousness, as would be the case in concurrent techniques.

In *retrospective probing* (RP), researchers prepare questions to ask participants once a task is complete, focusing on users' recent impressions as they performed certain actions. Meanwhile, in *retrospective think-aloud* (RTA), test moderators have participants retrace their steps after the interaction concludes and may even present test subjects with a transcript or recording of their conversation, asking follow-up questions to shed light on key moments.

To test Ask GeorgiaGov, we used retrospective probing exclusively with questions we asked users after the interaction was over, collecting insights about its performance. Our retrospective questionnaire consisted of three questions:

Facilitator: *"On a scale of 1–5, based on the scenario, was the information you received helpful? Why or why not?" (seeking quantitative and qualitative responses about content relevance and search issues)*

Facilitator: *"On a scale of 1–5, based on the scenario, was the content presented clearly and easy to follow? Why or why not?" (seeking quantitative and qualitative responses about voice content legibility and discoverability)*

Facilitator: *"What's the answer to the question you were tasked with asking?" (verifies that the user landed on correct voice content)*

Running voice usability tests

The core mission of a voice usability test is to gauge the *quality* of the interface along multiple dimensions. Because voice interfaces are more humanlike than other interfaces, even if an interaction was successful, the demeanor of the interface or the nature of the response may be grounds for a user to say it wasn't a good experience.

Your test procedure, written before the first round of tests, should outline not only the questionnaire you'll use for your retrospective testing, but also any other preemptive questions users need to be asked, such as demographic information or consent to record. Here are some of the initial questions we asked our usability test subjects after they completed our pre-screening questionnaire and sat down:

Facilitator: *"Is it okay if we record and transcribe?"*

Facilitator: *"On a scale of 1–5, what would you say your skill level with Amazon Alexa is, 1 being 'I've never used it before' and 5 being 'I use it for everything at home'?"*

Our test procedure also enumerated the processes by which we gave participants materials, like any needed guidance:

User is given a printout consisting of help text from the skill page on the Alexa Marketplace and a scenario to work with.

We also clearly described how users would proceed through scenarios.

Scenario: *"You're a registered nurse and you've just moved to Georgia, but you don't know if your license from your old state is still valid. Talk with Alexa to find out the information you need. At the end, ask for a phone number for more information."*

User is then told to start skill.

User: *"Alexa, ask GeorgiaGov."*

Our procedure ended with facilitators asking the three questions from the retrospective questionnaire, and then repeating the procedure to guide the user through two more scenarios, interceding if any issues arose during the test.

As you can see, robust test procedures ensure your evaluation is consistent and repeatable. Because our Alexa skill also involved the integration of discrete technologies such as a CMS, we encountered considerable challenges when it came to debugging, mostly involving hooking up Drupal's complex content services to Alexa's own complicated approach to handling data. We learned just how fundamental to the design process "test early, test often" truly is (http://bkaprt.com/vcu36/05-03).

Multiple rounds of usability testing will uncover problems in your interface that need addressing before deployment and launch. But even beyond this, there are a few other tasks I'd advise teams to perform before launching a voice interface.

JUST BEFORE LAUNCH

The *prerelease phase* differs from the iterative voice usability testing we've covered earlier in this chapter. We're now concerned with making sure that every piece of voice content is discoverable, that every point in the dialogue is accessible, and that the voice interface behaves in lockstep with call-flow diagrams.

- Tasks during the prerelease phase involve cross-functional teams and stakeholders, and we'll cover each in the coming sections:
- dialogue traversal testing (DTT)
- logging and analytics
- creating a custom dictionary (if needed)

Other tasks might be warranted in this phase, such as quality assurance (QA) testing to ensure software code performs as expected, and load testing to ensure many users can simultaneously access the interface. Typically, both of these are responsibilities of the underlying technology and the developer team.

Dialogue traversal testing

In *dialogue traversal testing* (DTT), someone on the team examines every nook and cranny of the interface to check if there's any point that could jeopardize the entire interaction. Confusing transitions, missing error-recovery strategies, or missing responses might surface in DTT.

Regardless of what kind of voice interface you're building, conducting a dialogue traversal test involves interacting with the fully functional voice interface. As Cohen, Giangola, and

Balogh write in *Voice User Interface Design*, "You should try silence to test no-speech timeouts. You should impose multiple successive errors within dialog [*sic*] states to ensure proper behavior" (http://bkaprt.com/vcu36/01-02).

It's crucial to traverse as many of the error trajectories associated with every dialogue state as possible, especially for the two most frequent errors voice interfaces need to handle: no-speech timeouts (when no speech was detected) and no-match errors (the system isn't equipped to handle the user's response).

There's no one right way to conduct dialogue traversal testing. An optimal *test script* ensures every dialogue state and every error-recovery strategy is visited. In some scenarios, especially open-ended prompts ("What can I do for you today?"), a complete round of DTT might be impossible due to the limitless range of possible user responses. For these interfaces, at minimum, this means ensuring that users can reach all basic functions, dialogue states, and happy paths.

In *Designing Voice User Interfaces*, Pearl suggests printing out the final call-flow diagram for your interface and testing each of the journeys or *traversal patterns* represented in the diagram, scrawling notes about problem areas in the interface in the process (**FIG 5.1**) (http://bkaprt.com/vcu36/03-02). Like running into every possible dead end in a labyrinth, DTTs are first and foremost about ensuring your users never get lost or waylaid at any critical point; they're not meant to be exhaustive, since that is usually unrealistic.

Logging and analytics

If your interface pulls content from a CMS, how often do 404 errors occur? Do users who ask about a particular topic run into more errors? Is there a point at which a handoff to a third party is timing out, returning nothing? *Logs and analytics*, which track users as well as their successes and failures across the entire interaction, help stakeholders measure how the quality of the interface evolves over time and inform debugging and maintenance later on.

Common metrics include the amount of time spent within the interface, task completion rates, dropout rates, barge-in

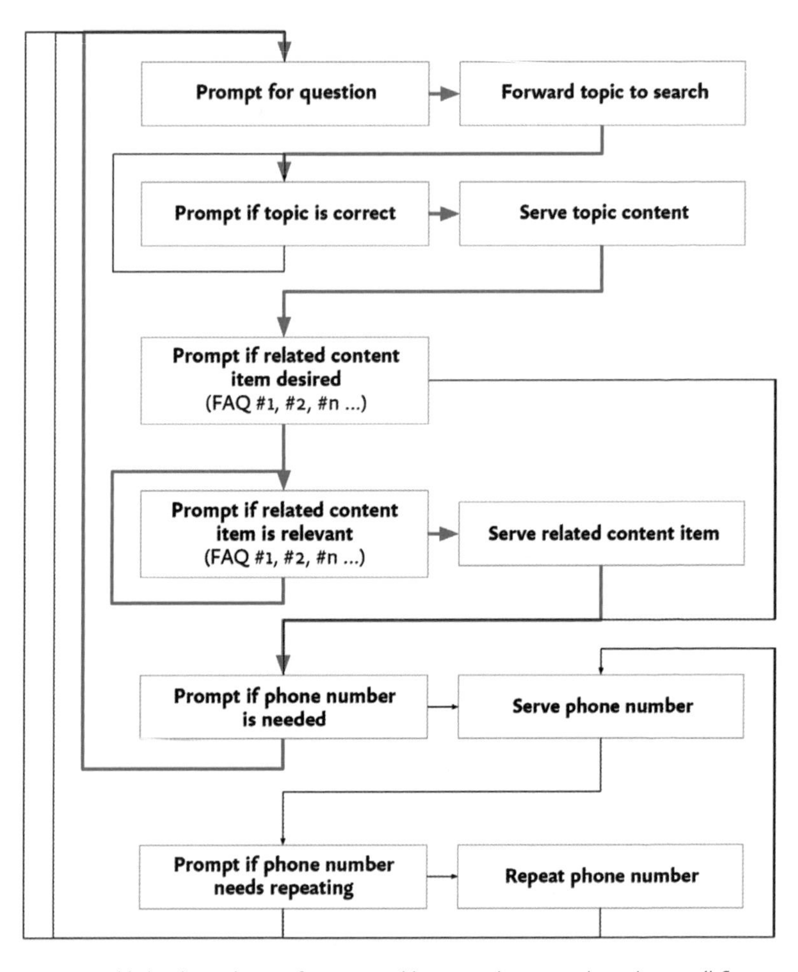

FIG 5.1: Highlighted in red, one of many possible traversal patterns through our call-flow diagram for Ask GeorgiaGov, in which we trigger a recursive behavior and decline a phone number.

(when a user interrupts an ongoing machine utterance) rates, and the frequency of no-speech timeouts and no-match errors. Each metric offers insight into how your initial users will judge the interface, allowing for quick calibration just after launch. For instance, a high barge-in rate suggests that users are growing frustrated too early and interrupting the interface before it's done talking.

For all of your analytics and logging to succeed, your voice interface needs to provide or export data to a format that stakeholders and maintainers alike can understand. Without a mechanism for logging, upon launch, you won't have a clue how your users are evaluating your voice content.

In the case of voice content, it's often the case that your CMS or database can capture logs in the same way they might track metrics for a website. In other cases, like when you depend on a database you don't control, you may need to forward the gathered data to a third-party system or dashboard. Logs help developers understand where things are going awry, but they are also critical for content strategists and editors who can glean insights on how to adjust and gauge the performance of their content in voice.

Logs should archive key information from interactions themselves to give both engineers and editors insight into user intents that result in more errors, machine prompts that confuse users, and content that requires reformulation. Your logs might encode information such as:

- the recognition result (how the system understood a user, with confidence scores; this might include audio recordings)
- the match result (what the system matched the user's utterance to)
- error states (no-speech and no-match errors)
- dialogue or decision state names (identifying where the user was in the interface)
- latency (to account for where delays may be occurring)

Because Ask GeorgiaGov forwards users' questions to the search service on the Georgia.gov website, we couldn't rely solely on the reports that Amazon Alexa itself provides developers and maintainers. In addition to Alexa's built-in logs, we built logs into the Drupal CMS that recorded not only errors but also events that Alexa couldn't handle solo, like search queries it needed to delegate. In the process, we were able to gather transcriptions of all search queries, queries that returned no results, and, most crucially, what content was pulled from Drupal into Alexa.

Conveniently for the Digital Services Georgia team, these logs and reports sat alongside the very same dashboards the editorial organization used to analyze the performance of the Georgia.gov website, allowing for quick and easy comparison between the two channels. If you're dealing with content delivered through multiple channels, consider what will make the lives of your content editors easiest.

Custom dictionaries

One crucial step before launch is to populate your *custom dictionary* if your framework permits it. In Amazon Alexa and Google Home, for instance, you can program certain terms and phrases that might be unfamiliar to the automated speech recognition capabilities within the hardware.

Say you're building an in-car navigation system for a specific city. Registering "Clemons Street" at a higher probability than "chemistry" can help Alexa more readily select the likelier result. In other cases, you may need to become a bit of a lexicographer and introduce terms to the interface's vocabulary that are absent from common usage. Some names, for example, may require a pronunciation that a speech synthesizer doesn't recognize from the natural language it encountered during its training (like "Houston Street" in Manhattan, which is not pronounced like the name of the Texan metropolis).

Check your dialogues for:

- brand names
- proper nouns
- neologisms
- jargon
- loanwords from other languages
- unusual pronunciations
- uncommon words
- homophones or similar-sounding terms needing disambiguation

Before deploying Ask GeorgiaGov, we discovered several issues that obligated the creation of a custom dictionary. Certain common terms in Georgia state law went unrecognized by Alexa, most notably *ad valorem tax,* a Latin term that refers to a tax based on the appraised value of a transaction or property. We added this and several dozen other terms to our custom dictionary. Depending on the kinds of terminology you deal with, you'll want to consider a dictionary, too, if your platform supports it.

VOICE CONTENT IN THE WILD

Ask GeorgiaGov was released on the Alexa Skills Marketplace in October 2017 to great fanfare by both the Acquia Labs innovation team and our client, Digital Services Georgia. We were overjoyed to receive feedback from real Georgians all over the Peach State who now had another means to reach their state government. And we had the evidence to back it up thanks to our usability test results and our logs and reports.

Eight months after Ask GeorgiaGov launched, we held a retrospective with the Digital Services Georgia team to leaf through the logs and review the reception among residents of Georgia. What we found shocked us all. As it turned out, there was a massive gap between the topics most searched for on the Georgia.gov website and those sought on the Ask GeorgiaGov interface. Users who preferred Alexa to acquire content searched most often for information regarding vehicle registration, driver's licenses, and the state sales tax. Though Ask GeorgiaGov has since been decommissioned, these results were deeply instructive for Georgia's subsequent forays into conversational interfaces, like the text-driven chatbot that carries the same name (http://bkaprt.com/vcu36/02-01).

Buoying all of us who built one of the first-ever content-focused Alexa skills, 79.2 percent of all user interactions over the preceding eight months had led to the delivery of a successful content response, a number that was striking due to the relative immaturity of Amazon Alexa at the time and the lack of informational voice interfaces in production. We also found that

71.2 percent of all interactions led users to request an agency phone number, validating our choice to broaden our thinking to devices beyond Alexa, like smartphones.

Buried deep in the reports, however, we found perplexing 404 errors citing a search term that appeared repeatedly in the logs as "Lawson's." It was an anomaly we couldn't unravel. After consulting the native Georgians around the table, we unearthed the culprit. As it turned out, one of our valued users, somewhere in Georgia, had been trying fruitlessly to get Alexa to understand her pronunciation of the word "license" in her native drawl (http://bkaprt.com/vcu36/05-03).

This heartwarming anecdote highlights an honest truth. Though testing early and often and preparing copious logs can bring an interface to within an inch of perfection, flaws still arise after launch, whether due to human error or, in this case, because of Alexa's own inability to understand our kaleidoscope of American English dialects. It seems we can rest assured that voice interfaces still have a way to go before they outwit us at our own game of human conversation.

6 THE FUTURE OF VOICE CONTENT

WE'RE RAPIDLY ENTERING a new era of voice interface design. Up to this point, this book has presented a contemporary account of how we design and build voice interfaces that may bear little relation to how they'll look—and sound—in the coming years. In the near future, how will new tools and technologies make voice interfaces not only more robust, but also more democratized for designers who neither write code nor want to deal with hardware?

Accelerating innovation in voice interfaces is rightfully leading designers and content strategists to pose key questions about where they fit in the picture, how lifelike speech synthesizers will truly perform, and whether approaches that aspire to authentic conversation will be available to anyone and everyone, not just rarefied corporations.

It's ironic that soon, many of the machine-centric tactics laid out in this text may give way to a much more human future. Design artifacts like sample dialogues and call-flow diagrams may yield the floor to intelligent algorithms and neural networks that can bamboozle people into thinking they are chatting with a fellow human being. But debate continues about whether voice users of today actually want a conversational partner who is indistinguishable from a real human being. Do we truly *want* to be hoodwinked? Will voice interfaces *actually* represent our multifaceted world?

After all, rooted in the notion of "perfectly human" voice interfaces is a brewing debate about how faithfully our voice interfaces ought to mirror our society. When the face of a voice interface is *necessarily* a human being, how do we resolve issues of representation and belonging so that users see and hear themselves reflected in the technology they use? As we move toward more people building voice interfaces on their own and narrow the gap between computers and ourselves, if we can repair our society's damaging biases that threaten that lofty vision in the process, the future of voice content will be bright.

THE DEMOCRATIZATION OF VOICE DESIGN

Since the advent of the first IVR systems, with their finicky hardware and dizzying requirements, the way designers and developers build voice interfaces has continued to evolve, lowering the barrier of entry to voice interface design for a wider audience. No longer do we need a postgraduate degree in computational linguistics or to understand the specifics of a call-routing system or VoiceXML to succeed.

A new competitive landscape in the conversational industry is bringing a medium formerly confined to large corporations down to the level of designers and developers working alone to craft voice interfaces. Thanks to these new technologies and frameworks that make designing and building voice interfaces easier than ever, the age of voice interfaces is not only available to some; it is now accessible to all of us.

But even though what-you-see-is-what-you-get (WYSIWYG) and drag-and-drop tools are beginning to surface for voice interfaces, many designers today still depend on developers. It's often still fundamentally the domain of developers to implement the voice applications designers and content strategists want to see. In some ways, the situation has changed little from the IVR-systems era.

Industry leaders are beginning to grasp in earnest the value of designer-friendly tooling for voice interfaces. Breaking this barrier for designers will democratize voice interface design in much the same way that no-code tools have revolutionized web design. In only a few short years, we may see even more no-code tools proliferate—joining the likes of Voiceflow, Botsociety, and Google's Dialogflow—that allow creatives to spearhead the process from start to finish, all without the aid of a developer. Soon, for the first time, voice interfaces will truly be buildable by anyone and everyone.

HOW VOICE WILL REFRAME CONTENT

For editorial teams, voice content will prove to be a compelling conduit to reach users beyond the website in the coming years. In addition to growing demands for voice-enabled content, we're also in the midst of an explosion in other channels for content delivery like mixed reality, wearables, and digital signage. Our page-bound content strategies and web-only content management systems will need a full refresh to keep up.

Today, we can no longer assume that our content will always display as a page. The era of macrocontent, the stuff of long-form websites, is giving way to more modular microcontent: atomic chunks of copy that resemble the sort of content we scroll through on social media, microblogging tools, and, of course, voice interfaces.

But how can organizations successfully transform legacy macrocontent into modern microcontent? After all, there isn't yet a good solution for reconciling content that needs to be rendered into distinct spoken and written forms. Without managing multiple versions (a potential maintenance headache), it's still difficult to imagine toggling between written text in a formal register and spoken text in a colloquial register on the fly.

While some content strategists might suggest the use of parallel renditions of text, one for a written website and the other for a spoken voice interface, this introduces an increased burden of upkeep for editorial teams. I'm interested in seeing more automated approaches become reality: a tool could traverse content items in a corpus and, based on certain settings, chunk that content into voice-ready pieces and even potentially convert formal, buttoned-up text into more fluid, conversational prose.

Some emerging services are headed in the right direction. Amazon Polly, for instance, renders written text into speech that sounds natural to the human ear and can integrate with CMSs like WordPress and Drupal to generate audio recordings of copy read aloud by configurable voices (http://bkaprt.com/vcu36/06-01). However, Polly doesn't yet support in-text transformations, like swapping the formal word *cannot* for the informal contraction *can't*.

A truly channel-agnostic solution would dynamically generate content on demand, optimized for voice and other contextless content experiences, all without demanding any manual modification of the original content source. Whatever innovations take hold, voice content won't just upend how we consume content; it'll also reinvent how we author, govern, and publish it.

INCLUSIVE VOICE CONTENT

More voice content will also benefit disabled people who employ assistive technologies such as screen readers to navigate web content, in addition to people who use mobility aids or can't use a keyboard and mouse. For Ask GeorgiaGov, for instance, we discovered that many elderly and disabled Georgians could acquire information far more quickly from our voice content than by traveling in person to or calling an agency office. If this was our experience, the implications of voice content for accessibility and inclusion could be staggering.

For Blind users, screen readers leave much to be desired, because the visual nature of the web limits efficient aural navigation. Interfaces that present voice content offer a chance for Blind people to interact with content using conversational structures, which are far more understandable than a screen reader's verbose translation of what a web page looks like. Never before have other user interfaces mounted a serious challenge to screen readers as the primary means to access web content for Blind users. Voice interfaces slinging voice content could conceivably even supersede screen readers as a whole.

When we extricate ourselves from a single-channel bias and evaluate how different users interact in distinct ways with content on a range of devices, we can better meet their needs and make all access to content more equitable. Delivering content through voice could support someone with chronic migraines who can't stare at a screen for more than a few minutes, or someone with Parkinson's disease who prefers to speak instead of moving a mouse. Offering multiple keys to unlock content gives people with different lived experiences more choice, more flexibility, and more equity.

But throughout this book, we've identified a problem: What's the use of providing voice content if the conversations we have with an interface are stilted and repetitive? If we truly want to meet every user where they are, shouldn't we foster more lifelike human conversation?

CONVERSATION-CENTRIC DESIGN

There's something eerily soulless about reducing our entire capacity for conversation to two seemingly ill-fitting types: transactional and informational voice interactions. After all, among the acknowledged pitfalls of voice interfaces is their uncanny aloofness. User experience researchers Robert Moore and Raphael Arar contend in *Studies in Conversational UX Design* that the voice interfaces of today aren't adequate for true conversation, precisely because of this rigidity:

> *Creating a user interface that approximates [natural conversation] requires modeling the patterns of human conversation, either through manual design or machine learning. Rather than avoiding this complexity, by producing simplistic interactions, we should embrace the complexity of natural conversation because it mirrors the complexity of the world. The result will be machines that we can talk to in a natural way, instead of merely machines that we can interact with through unnatural uses of natural language.* (http://bkaprt.com/vcu36/03-03)

According to this argument, voice interactions that feel natural are more faithful to the sort of organic and meandering conversation we might have in passing with a cashier at the grocery store—at times transactional, at times informational, and at times prosocial.

Most voice interfaces up until now have been solely transactional or informational, with limited capacity for idle chatter beyond some onboarding and feedback. But today, voice assistants like Amazon Alexa and Google Home compete with each other on the basis of their affinity for human conversation. After all, as conversational designer Deborah Dahl emphasized at Mobile Voice 2016, a voice assistant "just doing a web search doesn't show understanding" (http://bkaprt.com/vcu36/06-02).

Sadly, the utopian ideal of *conversation-centric design*—equipping a voice interface with the capacity for truly extemporaneous dialogue—is still a long way away. Attractive, conversation-centric design requires budgets that most organizations simply don't have. Without the full force of technical teams exploring the latest and greatest in low-level speech technologies, which remain luxuries only behemoths like Amazon, Apple, and Google can afford, a truly conversation-centric interface remains a pipe dream for most designers.

Voice interfaces occupy a new niche in our growing collection of artificially learned interfaces, right next to keyboards, mice, and game controllers. We've gotten used to their mechanical awkwardness and cold repetitiveness; and in the process, we've evolved a new category of artificial interactions to acquire and rehearse. Learning to type quickly on a keyboard, then, exercises much the same kind of muscle memory as learning to have effective voice interactions with Alexa or Google Home, but those rehearsed behaviors don't necessarily pass as true conversation.

This highlights the paradox of conversation-centric design. Designing voice interfaces that are more stilted allows us to limit the interface's responsibilities to what's within reason, like delivering content related to a single topic. It keeps our scope tight. But conversation-centric design to its fullest extent means washing away those boundaries and meeting the expectations humans have of a normal conversation. True conversation-centric interfaces can answer any conceivable question instead of only handling preprogrammed transactional or informational use cases.

That said, while the ideal of conversation-centric design poses additional problems when it comes to user expectations, that doesn't mean we can't aspire toward conversation-centricity with custom-built informational and transactional interactions—and even a smidgen of prosocial small talk, even if it's on the kitschy side—that coalesce toward the user's happiest path. Moreover, conversation-centric design is a fundamental component of how technologists see voice interfaces innovating in the coming years.

THE CONVERSATIONAL SINGULARITY

Perhaps the most urgent question for the voice industry at large is the one whose answer will render this book obsolete and relegate it to bookstore bargain bins: When will the manual design strategies we've covered in this book, like sample dialogues and call flows, give way to truly conversation-centric design?

After all, the mission of voice assistants like Alexa and Google Home is to make *all* content across the web available to voice users, not just a carefully curated subset of it siloed away on a site somewhere. Though we remain far from truly natural conversation, substantial investments by the likes of Amazon, Apple, Google, and IBM are well underway to outperform the highly structured interactions that voice users have with their devices today using custom-built skills.

Slowly but surely, we're making our way toward what Mark Curtis calls the *conversational singularity*, a moment when the frictions between humans and conversational interfaces vanish because machines are capable of having bona fide organic conversation at last (http://bkaprt.com/vcu36/06-03). Dependent on advancements in natural language processing, natural language generation, and speech synthesis, the conversational singularity will render our carefully constructed, custom-built voice interfaces obsolete—but will swing the door open to new promise for voice experiences.

The conversational singularity is compelling for futurists, because it portends a future where voice interfaces display what Harris calls *habitability*, which he defines in *Voice Interaction Design* as characterizing "a system the user can inhabit." Rather than constraining the user like gutter guards in a bowling alley, Harris suggests voice interfaces could do the work to accommodate the user ergonomically and adapt intelligently to whatever they have in mind (http://bkaprt.com/vcu36/02-02).

Right now, we're in that strange transitional time where anyone can architect voice interfaces, but we must still program them in ways diametrically opposed to the ideal of conversation-centric design. Soon, however, the corporations that gave us Siri, Cortana, and Alexa may reach capabilities that go well beyond all custom-built voice interfaces on earth. This outcome may sound like a happy ending, but it is in fact a dangerous proposition.

One of the less desirable effects of the conversational singularity could be the continued concentration of the reins and levers governing conversational technology in the hands of the wealthy and privileged few. It could also lead to mass layoffs of customer service agents and call center staffers around the world, many of them in low- and middle-income countries. This leads us to our final topic: inclusion and representation in the voice world itself.

IDENTITY AND INTRINSIC BIAS IN VOICE

When you hear Alexa, Cortana, or Siri speaking, who is the person you picture in your mind?

We eagerly anthropomorphize the machines we have conversations with, even though machines couldn't care less about how they are personified. Despite the fact that they are entirely digital automata, we ascribe traits to them and name our voice assistants *Alexa* and *Siri*. We treat such "executive assistants" as cisgender white women, despite the misogyny and racism inherent in such characterizations—not to mention the fact that faceless voice assistants bear no resemblance to a real person.

A human identity can give life to normally austere interfaces like IVR systems and voice assistants, but it comes at a cost. The sexism in our society that portrays executive assistants only as secretarial women pervades every aspect of voice assistants, including how we talk to them. Deeply held biases prejudice us in favor of one type of voice assistant over another. The attributes we bestow upon speech synthesizers to establish identity may in fact worsen the systemic oppression that many of us face on a daily basis in society.

Because those tasked with building the first IVR systems were generally straight cisgender white men, it's no surprise that most voice assistants are, to the user's ears, straight cisgender white women who speak in a General American dialect. There's seemingly no space for dialects like Indian English and African-American Vernacular English (AAVE), immigrants who speak English as a second language, and queer people who code-switch between LGBTQ+ and straight-passing modes of speech.

There are some tentative emerging steps in the right direction. In-car navigation applications like Waze now permit users to upload their own short audio clips to replace the default voice recordings that pepper rides with various versions of "Keep left" and "Accident reported ahead." Once, while I was in a Lyft humming down an expressway, my driver shared how he'd enlisted his school-age daughter to record her own voice. "I wanted to have her with me on every ride," he told me proudly as she rattled off directions.

I'm eager to see this level of voice customization, or at least a broad selection of possible voices to choose from, become available for speech synthesizers in voice assistants. Not only would it pave the way for better representation of diverse dialects and lived experiences; it would also facilitate more inclusive and equitable forms of voice interface design.

Our voice interfaces are capable of confirming or challenging our deeply held biases about society. Today's audiences for voice interfaces are increasingly attuned to the need for representation of the marginalized and underrepresented, which might manifest in voice interactions through bilingual or dialectal code-switching, colloquialisms used among communities of color, and speech synthesizer customization. Because voice interfaces are the most humanlike of all digital experiences (a refrain throughout this book), we must respect—and celebrate—the humanity of those we aim to reach.

REPRESENTATION MATTERS

Inclusion in voice design is not merely about accessible alternatives to screen readers or hearing a variety of voices. It's also about authentic representation. After all, the digital identities we give voice interfaces can introduce human problems that machines are not coded to consider.

In the realm of voice, interfaces are no longer just machines—we see them, for better or worse, as people with their own fully formed identities. In order for voice content to become truly inclusive of people having diverse lived experiences, creators of voice interfaces must grapple with the systemic issues that surface. My sincere hope is that corporations like Microsoft, Apple, Google, and Amazon instigate top-down approaches to improve inclusion and customization in speech synthesizers.

In proprietary software, however, this is unlikely. So, at the same time, those responsible for the foundations of speech synthesis all voice designers rely on should open avenues for others to create vocal patterns and speech styles that more accurately represent the real world. Such a grassroots, bottom-up solution would require speech synthesis technologies to be open-sourced, and the potential progress could reinvent representation in voice interfaces forever.

I hope for a not-too-distant future where we can configure the voices we hear on Alexa like we do on Waze, where those who engage in bilingual code-switching can hear those same toggles represented on Google Home, and where Black trans women and other multiply marginalized groups can hear someone from their own communities telling them the baggage carousel for their flight, the closing time of continental breakfast, and the best way to get a small business loan.

The golden age of voice content is rapidly approaching, but we remain in the early stages of a mass migration of formerly web-only content to newly voice-enabled content. We must not let it fall victim to the vagaries of human society that our machines simply couldn't care less about, like the inherent biases—racism and ableism, misogyny and misogynoir, homophobia and transphobia—that continue to silence the people we serve.

We've had a long-running conversation throughout this book about the amazing things that can happen when we give our copy, long trapped on the web, a second life as voice content. Now, it's time for us to work toward the next and far more important milestone as technologists. It's your turn to take up the challenge. Just imagine what we can accomplish and invent if we can restore to those who have long been oppressed and underrepresented in our society a voice—one that each and every one of us can truly hear ourselves in, and call our own.

ACKNOWLEDGMENTS

IT TAKES A VILLAGE to write a book, especially for a topic as broad, ever changing, and vibrant as voice interfaces. The initial ideas and rough outlines for *Voice Content and Usability* began to take shape soon after the launch of Ask GeorgiaGov in 2017 and the publication of my debut book *Decoupled Drupal in Practice* in 2018, but it wasn't until 2020 that inkblots finally made it onto paper.

There are far too many people to thank within the limited space of these pages, but I wish to highlight several folx in no particular order who have been instrumental in forming the ideas underpinning this book and the creation of this new contribution to voice literature. Without these individuals and their unyielding support, this book never would have made it past the drawing board.

First and foremost, I wish to thank Chris Hamper, the architect and developer behind the Ask GeorgiaGov Alexa skill, who worked side by side with me at Acquia Labs (Acquia's innovation arm) on pilots ranging from voice to beacons to augmented reality to digital signage for some of the best-known clients in the world, including Georgia and Nestlé Purina. In the spirit of open source, Chris contributed extensively to the Alexa Drupal module originally developed by Jakub Suchy, the nucleus of Ask GeorgiaGov now used by many Drupal sites in production. Thanks also to Dries Buytaert, Drew Robertson, Leah Magee, Alena "ASH" Heath, and the former Office of the CTO (OCTO) at Acquia, all of whom supported our harebrained experiments at Acquia Labs in various ways.

Second, we at Acquia Labs could not have found a better partner than the incredibly knowledgeable and compassionate team at Digital Services Georgia, who enthusiastically worked with us to design and build the first-ever voice interface for residents of Georgia, despite challenges like an unprecedented content auditing process for voice. Thank you to Nikhil Deshpande, a dear friend, for his confidence and trust in us despite the risks; and to the entire Digital Services Georgia team who collaborated with us over the course of 2017, including cherished team members Kendra Skeene, Jenna Tollerson, Rachel Hart, and Donna Sumner.

Third, this book was the culmination of a unique series of events that connected me with the A Book Apart team. Eric Meyer, Toby Malina, and Jeffrey Zeldman discovered my Frontend United Utrecht 2018 session about designing content-driven conversational interfaces and invited me to speak on the subject at An Event Apart. Though I was only able to present at one conference edition before the pandemic took hold, this spark ultimately resulted in this new contribution to the literature becoming a tangible reality. My sincere gratitude goes to Jeffrey Zeldman in particular for generously giving his personal time in February 2020 to support me on this journey and to offer advice I continue to hold very close to my heart today.

I also wish to thank the *A List Apart* team, who were an absolute pleasure to work with as I authored and published "Usability Testing for Voice Content," the first distillation of some of the key ideas in this book. Fond thanks to Aaron Gustafson, my editor Desirée Zamora García, and producer Brandon Gregory for their support.

I'd be remiss not to give ample space to the entire A Book Apart team, whose editors and staff have been an utmost joy to work with. I've never undertaken a writing process quite like this in these extraordinary circumstances, but my supporters at A Book Apart have been steadfast resources throughout the ups and downs. My dear thanks to my editors Lisa Maria Marquis (whom I credit also with igniting my newfound passion for everything *Star Trek*) and Sally Kerrigan (whose feedback has been both instrumental and deeply thought-provoking), and a big thank-you also to Caren Litherland, Katel LeDû, and Jason Santa Maria.

Working with both voice interfaces and content management systems can be complex, and I wish to thank Gaurav Mishra, a dear friend and longtime booster of my work, for donating his time as a technical editor and his expertise as a subject matter expert who lives and breathes voice technologies and CMSs like Drupal each and every day.

Last but not least, I wish to thank all of my friends across the world, my fellow flinty organizers at Decoupled Days, and colleagues past and present at Time Inc., Acquia, Gatsby, and Oracle who have been nothing but encouraging throughout this project. My former Gatsby colleagues Marcy Sutton and Jason Lengstorf have not only been deep inspirations for me in their own right as technologists; they also boosted this endeavor at a time when I wasn't sure I'd be able to find an interested publisher. Thanks also to my Oracle colleague Girish Bettadpur, a mentor who has been enthusiastic in his backing of me both personally and professionally.

As a parting note, I want to offer my personal words of love and support to all marginalized folx who have yet to find their own voice in the discourse due to systemic oppression and deep-seated bigotry. The world needs voices like yours in order to serve our beautifully diverse society fairly and equitably. I hope my experience inspires you to contribute your own invaluable insight. This book is for you.

RESOURCES

VOICE INTERFACE DESIGN is a vast and burgeoning field. It encompasses many disciplines from linguistics and rhetoric, to hardware and software engineering, to accessibility and usability. This new contribution stands on the shoulders of literature from the IVR-systems era onward to today's voice assistants. Here are some of the books, articles, and resources that've helped me make sense of the chaos.

Voice interface design

- *Voice User Interface Design* by Michael Cohen, James Giangola, and Jennifer Balogh is perhaps the most thorough and insightful bible on voice interface design in existence today. Though its prescriptions can be narrow and its case studies light on the details, its breadth means it's a great overview for anyone getting started. If you only read two books from this list, make this one of them (http://bkaprt.com/vcu36/01-02).
- The other one to read is *Designing Voice User Interfaces: Principles of Conversational Experiences* by Cathy Pearl, which covers foundational principles of voice interface design, with roots in the IVR-systems era (http://bkaprt.com/vcu36/03-02).
- *Voice Interaction Design* by Randy Harris is a bit of a dense read but covers all the mechanics and mechanisms by which voice interfaces do their magic, including everything from call-flow diagrams to the concept of habitability (http://bkaprt.com/vcu36/02-02).
- Tools abound for crafting dialogues and planning call flows. Some voice designers use Celtx, a tool for writing screenplays, but virtually any word processor can format sample dialogues. For call-flow diagramming, tools like OmniGraffle and Whimsical are great, but I use the drawing feature in Google Docs or Google Slides.

- By Susan L. Hura, Sarah Turney, Lizanne Kaiser, Todd Chapin, Fran McTernan, and Caroline Nelson, "The Evolution of VUI Design Methodology" is a comprehensive look at the evolution of voice interface design processes and tools over time (http://bkaprt.com/vcu36/07-01, PDF).
- *Practical Speech User Interface Design* by James Lewis is lengthy and heavy on the implementation side of things, but it includes a variety of lessons and directions to pursue, especially when it comes to handling onboarding, prompts, and intent identification (http://bkaprt.com/vcu36/07-02).

Conversational design

- Erika Hall's *Conversational Design* isn't just about how conversation works; it's also about how to *think* about conversation in the right ways and at the right scope. Her recommendations for approaching conversational design resonate everywhere in voice. Hall also has an *A List Apart* article with the same title (http://bkaprt.com/vcu36/07-03).
- *The Conversational Interface: Talking to Smart Devices* by Michael McTear, Zoraida Callejas, and David Griol is a page-turner of an introduction to conversational interfaces writ large and especially how human conversation translates into synthesized speech (http://bkaprt.com/vcu36/01-01).
- *Studies in Conversational UX Design*, edited by Robert Moore, Margaret Szymanski, Raphael Arar, and Guang-Jie Ren, is an academic collection of papers that explore both real-world case studies in conversational usability and futurist aspirations like conversation-centric design (http://bkaprt.com/vcu36/03-03).
- Books on chatbots tend to be more developer-oriented, but they have a lot to tell us in the voice world. *Designing Bots: Creating Conversational Experiences* by Amir Shevat and *Hands-on Chatbots and Conversational UI Design* by Srini Janarthanam are both excellent starts to begin drawing parallels (http://bkaprt.com/vcu36/03-01, http://bkaprt.com/vcu36/07-04).

- For more insights from chatbot designers, I highly recommend Eunji Seo's "19 Best UX Practices for Building Chatbots" and "11 More Best UX Practices for Building Chatbots" from *Chatbots Magazine* as well as Yogesh Moorwani's "Designing Chatbots" from *UX Collective* (http://bkaprt.com/vcu36/07-05, http://bkaprt.com/vcu36/04-02, http://bkaprt.com/vcu36/07-06).
- From *A List Apart,* Matty Mariansky's articles "All Talk and No Buttons: The Conversational UI" and "Designing the Conversational UI" do a great job of introducing some of the main things to think about when designing for conversation (http://bkaprt.com/vcu36/07-07, http://bkaprt.com/vcu36/07-08).
- Joscelin Cooper's "Do Androids Dream in Free Verse?" from *A List Apart* is an excellent treatise on writing good interface text for voice interfaces (http://bkaprt.com/vcu36/01-08).

Voice content

- Three *A List Apart* articles tackle the idea of information transmitted through voice interactions from compelling angles. "Conversational Semantics" by Aaron Gustafson and "Conversations with Robots" by Andy Fitzgerald examine what can be done from the perspective of web technologies, while Caroline Roberts's "The FAQ as Advice Column" examines how FAQs can manifest as voice-friendly content (http://bkaprt.com/vcu36/01-06, http://bkaprt.com/vcu36/07-09, http://bkaprt.com/vcu36/07-10).
- Our work on Ask GeorgiaGov at Acquia Labs led to case studies by both Acquia Labs and Digital Services Georgia, as well as a VOICE Summit presentation about the project. It was also the subject of my session at Frontend United 2018 in Utrecht (http://bkaprt.com/vcu36/07-11; http://bkaprt.com/vcu36/07-12; http://bkaprt.com/vcu36/07-13, video; http://bkaprt.com/vcu36/07-14, video).

Voice usability

- Susan L. Hura, one of the seminal figures in voice interface design, has an excellent overview of some of the problems of voice usability testing in "Usability Testing of Spoken Conversational Systems" (http://bkaprt.com/vcu36/01-03, PDF).
- My *A List Apart* article "Usability Testing for Voice Content" delves deeper into how the unique problems of voice and voice content demand equally unique solutions when it comes to content strategy and usability testing (http://bkaprt.com/vcu36/05-03).

REFERENCES

Shortened URLs are numbered sequentially; the related long URLs are listed below for reference.

Chapter 1

01-01 https://www.springer.com/us/book/9783319329659

01-02 https://www.indiebound.org/book/9780321185761

01-03 https://uxpajournal.org/wp-content/uploads/sites/8/pdf/JUS_Hura_August2017.pdf

01-04 http://web.archive.org/web/20070928210916/http://www.visugate.biz/bjvi/1986/autumn1986.html#RCEVH

01-05 https://cs.stanford.edu/people/eroberts/courses/soco/projects/2005-06/accessibility/firstwave.html

01-06 https://alistapart.com/article/conversational-semantics/

01-07 https://www.wired.com/2016/03/war-stories-my-journey-from-blindness-to-building-a-fully-conversational-user-interface/

01-08 https://www.scientificamerican.com/article/the-semantic-web/

01-09 https://anildash.com/2002/11/13/introducing-microcontent-client/

01-10 https://alistapart.com/article/do-androids-dream-in-free-verse/

Chapter 2

02-01 https://georgia.gov/chat

02-02 https://www.elsevier.com/books/voice-interaction-design/harris/978-1-55860-768-2

02-03 https://digitalservices.georgia.gov/blog-post/2017-10-24/write-alexa-even-if-you-just-have-website

02-04 https://bigmedium.com/jhc/prez/mobile-myths.pdf

Chapter 3

03-01 https://www.oreilly.com/library/view/designing-bots/9781491974810/

03-02 https://www.cathypearl.com/book

03-03 https://www.springer.com/us/book/9783319955780

Chapter 4

04-01 https://oregonstate.edu/trees/dichotomous_key.html

04-02 https://chatbotsmagazine.com/11-more-best-ux-practices-for-building-chatbots-67362d1104d9

04-03 https://sensible.com/dont-make-me-think/

Chapter 5

05-01 https://uxdesign.cc/tips-for-accessibility-in-conversational-interfaces-8e11c58b31f6

05-02 https://www.usability.gov/how-to-and-tools/methods/running-usability-tests.html

05-03 https://alistapart.com/article/usability-testing-for-voice-content/

Chapter 6

06-01 https://docs.aws.amazon.com/polly/latest/dg/what-is.html

06-02 https://www.uxmatters.com/mt/archives/2018/01/designing-voice-user-interfaces.php

06-03 https://www.fjordnet.com/conversations/were-moving-to-a-conversational-singularity-mark-curtis-at-meet-the-media-guru/

Resources

07-01 http://acixd.org/wp-content/uploads/2018/10/DesignMethodology.pdf

07-02 https://www.routledge.com/Practical-Speech-User-Interface-Design/Lewis/p/book/9781439815847

07-03 https://alistapart.com/article/conversational-design/

07-04 https://www.packtpub.com/product/hands-on-chatbots-and-conversational-ui-development/9781788294669

07-05 https://chatbotsmagazine.com/19-best-practices-for-building-chatbots-3c46274501b2

07-06 https://uxdesign.cc/how-to-design-a-robust-chatbot-interaction-8bb6dfae34fb

07-07 https://alistapart.com/article/all-talk-and-no-buttons-the-conversational-ui/

07-08 https://alistapart.com/article/designing-the-conversational-ui/

07-09 https://alistapart.com/article/conversations-with-robots/

07-10 https://alistapart.com/article/the-faq-as-advice-column/

07-11 https://www.acquia.com/blog/ask-georgiagov-alexa-skill-citizens-georgia-acquia-labs/12/10/2017/3312516

07-12 https://digitalservices.georgia.gov/blog-post/2017-10-12/alexa-ask-georgiagov

07-13 https://www.youtube.com/watch?v=Gt6J_N9HmnQ

07-14 https://www.youtube.com/watch?v=kAy6AP-MZ4o

INDEX

ABOUT A BOOK APART

We cover the emerging and essential topics in web design and development with style, clarity, and above all, brevity—because working designer-developers can't afford to waste time.

COLOPHON

The text is set in FF Yoga and its companion, FF Yoga Sans, both by Xavier Dupré. Headlines and cover are set in Titling Gothic by David Berlow.

This book was printed in the United States using FSC certified papers.